T0175106

Routledge Introductions to Development
Series Editors:
John Bale and David Drakakis-Smith

Pacific Asia

Pacific Asia – from Burma to Papua New Guinea to Japan – is a new name for a region which no longer looks to the West for geographical definition. It is the most dynamic and productive region in the developing world, the result of an economic explosion fuelled by the industrial activity of the 'four little tigers' of Hong Kong, Singapore, South Korea and Taiwan. Pacific Asia is where the Green Revolution began, where there are more women in factory work than anywhere else; the region is also the most predominantly socialist in the Third World.

Pacific Asia reviews the historical development of the region and considers its present position in the global context. This invaluable introduction places general issues – resources, urbanization, ethnicity, population, gender, rural development and industrialization – common to all the countries of the region in local context. The author argues that these issues must be examined before successful policy responses can be determined.

A volume in the **Routledge Introductions to Development** series edited by John Bale and David Drakakis-Smith.

In the same series

John Cole
Development and Underdevelopment
A profile of the Third World

David Drakakis-Smith
The Third World City

Avijit Gupta
Ecology and Development in the Third World

John Lea
Tourism and Development in the Third World

John Soussan
Primary Resources and Energy in the Third World

Chris Dixon
Rural Development in the Third World

Alan Gilbert
Latin America

Janet Henshall Momsen
Women and Development in the Third World

Allan and Anne Findley
Population and Development in the Third World

Rajesh Chandra
Industrialization and Development in the Third World

Tony Binns
Tropical Africa

Jennifer A. Elliott
An Introduction to Sustainable Development

Allan Findlay
The Arab World

Mike Parnwell
Population Movements and the Third World

George Cho
Trade, Aid and Global Interdependence

David Simon
Transport and Development in the Third World

David Drakakis-Smith

Pacific Asia

London and New York

First published 1992
by Routledge
11 New Fetter Lane, London EC4P 4EE

Simultaneously published in the USA and Canada
by Routledge
29 West 35th Street, New York, NY 10001

Reprinted 1996

Transferred to Digital Printing 2004

Typeset by J&L Composition Ltd, Filey, North Yorkshire

British Library Cataloguing in Publication Data
Drakakis-Smith, D. W. (David William) *1942–*
Pacific Asia. – (Routledge introductions to development)
1. South-East Asia. Economic conditions
I. Title
330.959

Library of Congress Cataloguing in Publication Data
Drakakis-Smith, D. W.
Pacific Asia/David Drakakis-Smith.
(Routledge introductions to development)
Includes bibliographical references and index.
1. Pacific Area–Economic policy.   2. Pacific Area–Economic conditions.   3. Asia–
Economic policy.   4. Asia–Economic conditions–1945–   I. Title.   II. Series.
HC681.D73 1991
338.95–dc20   91-10872

ISBN 0–415–06985–8

Contents

Plates

Figures

Tables

Acknowledgements

As always a book is the product of many people's labour. I am particularly grateful to Pauline Jones and May Bowers who typed the manuscript, to Andrew Lawrence for his cartographic skills and to Maralyn Beech for her photographic work.

I would also like to thank the governing body of St John's College, Oxford, whose generous offer of a visiting scholarship enabled me to complete most of the manuscript during a summer of intensive work in stimulating surroundings.

The book is dedicated to my wife Angela and my children Chloe and Emmanuel. As any author knows, it is one's immediate family that shares all the burdens of producing a book but few of the credits.

David Drakakis-Smith

1
Introduction: the regional character

Defining Pacific Asia

The term Pacific Asia is relatively new and will be used in this book to refer to the developing countries identified in Figure 1.1. Japan and China will largely be excluded from the discussion: the former because it is now one of the most advanced nations in the world and no longer forms part of the developing countries of the region; the latter because it is so large and complex that another book in this series (forthcoming) is devoted to it. Clearly, however, a regional text on Pacific Asia cannot ignore two important states and the later chapters, particularly Chapter 2, give full recognition to their role in shaping the character of Pacific Asia.

In the past the region covered by this volume has been known, in part or as a whole, by a variety of descriptive terms, viz. East of Suez, the Far East, East Asia, Southeast Asia and so on. The emergence of the term Pacific Asia during the 1980s is a belated recognition of the fact that the global orientation of the countries in the region has changed. Until the Second World War this was a region dominated by colonial interests, most of which were still European. By the 1950s, however, that European influence had ended and the dominant power in the region was the United States.

Economic exploitation and political pressure during the 1950s and 1960s were, therefore, predominantly from across the Pacific. Since the 1970s, however, Japan has emerged as a leading political and economic nation, successfully challenging the United States for power and

Figure 1.1 Pacific Asia: political units covered in this text

influence amongst the capitalist economies in the region. In short, the region now looks east and has become much more conscious of its geographical position in relation to the Pacific rather than to Europe. Terms such as 'Near' or 'Far' East are thus anachronisms since they described geographical location with regard to the European colonial powers. We are about to enter the Pacific century and this part of Asia has renamed itself in preparation.

Clearly, we cannot undertake a study of Pacific Asia without emphasizing its global ties, both past and present, and the role it plays in the world economy. And yet the region is not uniform in character. There are sharp political divisions between socialist and capitalist states, whilst massive socio-economic contrasts also exist between the intensive industrial societies of Hong Kong or Singapore and the impoverished upland regions of Southeast Asia. Each country has developed in its own fashion, has its own set of resources, has reacted differently to colonialism, has been drawn into the world economy in varied ways, and has its own preoccupations in terms of development.

Yet this immensely varied set of nations does form part of the global economy and shares a common set of developmental problems – the legacies of colonialism, uneven regional development, rapid urbanization, the need for economic diversification and so on.

Given such circumstances it would be futile either to discuss each of these issues for the region as a whole, since differences are too marked, or to describe each country in turn, using an old-fashioned regional approach which runs through physical features, climate, resources, agriculture, etc. Instead, the general background to common contemporary developmental issues will first be examined (the global setting and internal diversity in this chapter; the historical evolution of the present situation in Chapter 2). This will be followed by discussion of selected major issues in the setting of just one or two countries, in order to illustrate how common regional and global development problems must be placed in both a national and local context before the real policy issues can be discussed.

Pacific Asia in its global setting

Trade

In the colonial period, the principal trade links between Pacific Asia and the rest of the world, primarily Europe, were essentially those of commodity exports (industrial raw materials and some foodstuffs) and manufactured imports. The present world trading system maintains vestiges of that era in the sense that many countries in Pacific Asia still rely on primary commodity exports for much of their foreign exchange earnings. However, as Figure 1.2 also indicates, manufacturing exports are now important throughout the region. Figure 1.2 does not reveal the complete picture, as it does not include earnings from the export of services: this is becoming increasingly important for the more advanced states in the region, such as Singapore and Hong Kong (see Chapter 7).

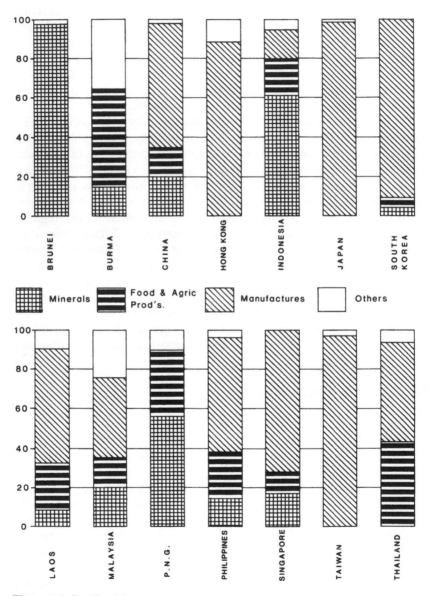

Figure 1.2 Pacific Asia: structure of exports

In general, the region's trade has increased steadily over the last 25 years, although it has slowed down during the recession of the 1980s, increasing threefold between the mid-1970s and mid-1980s. Overall, Pacific Asian trade (excluding Japan and China) now accounts for almost 10 per cent of all Third World trade.

Figures such as these illustrate the nature of Pacific Asia's growing integration with the world economy but, of course, there are considerable variations within the region. Burma, for example, has withdrawn into itself and its overall trading activity has steadily fallen over the years. Moreover, although the more industrialized states may cluster at the top of the aggregate trading lists, it would be unwise to forget that the sheer size of a nation will also be important in trading activity. Thus, although Indonesia's per capita prosperity is less than a third of that of the neighbouring state of Malaysia, its exports are larger.

Given the attempts of most countries in Pacific Asia to diversify from primary exports into manufacturing, it is not surprising that the region's imports are dominated by two categories – fuel and capital equipment. With regard to the latter, most of the countries in Pacific Asia spend at least one-quarter to one-third of their import bill on production machinery, and this includes the more industrialized states; some spend an even higher proportion. With regard to fuel, the region is sharply polarized between the energy rich, such as Indonesia and Malaysia, and the energy poor, which covers all of the major industrial exporters (including Japan). The result is a net flow of energy across the region from southeast to northwest.

But perhaps the most revealing aspect of regional trends in trade is not so much the overall level or even the principal type of exports and imports, but the direction of trading activity. Figure 1.3 reveals clearly how the formerly dominant European destinations have been overwhelmingly replaced by Pacific links. For the most part this involves Japan and the United States, with the former being the leading trading focus for the region since the mid-1970s. To be sure, some of the socialist states trade mainly with the communist bloc but even these have witnessed a growth in their trading activity with Japan. A slightly different picture emerges if the volume of trade is broken down into specific categories. For example, despite the overall trading dominance of Japan, the United States is still by far the most important market for the export of manufactured goods.

The corollary of much of this trading activity, particularly in exports, is the inward flow of investment into the region. All types of financial

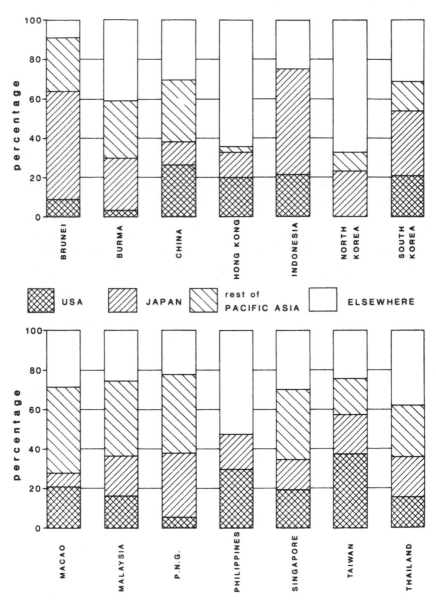

Figure 1.3 Pacific Asia: major trading partners

flows into Pacific Asia have increased over the last 20 years but, in general, private flows have increased far more rapidly than public or government flows. As a result, foreign investment has become a much more prominent proportion of total investment, particularly in those countries favoured by the international investors. Singapore, in particular, has been a magnet for such financial flows, so much so that half of its gross domestic product is funded this way.

The less desirable consequence of overseas loans, rather than investment, is debt and many of the countries in the Pacific Asian region face a mounting debt crisis as a result of borrowing during times of expansion and not being able to service the interest during the lean years of the 1980s (Figure 1.4).

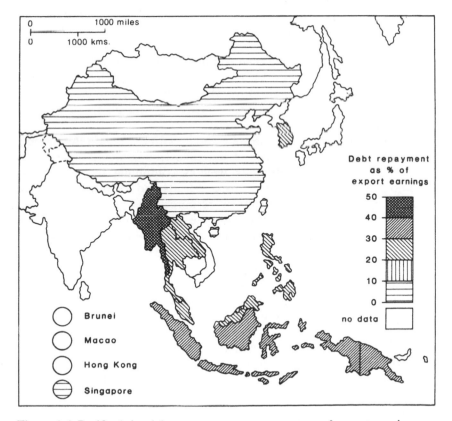

Figure 1.4 Pacific Asia: debt repayment as a percentage of export earnings

Resources

Clearly the investment and expansion summarized in the previous section is attracted by the resources of the region. The region is reasonably well endowed with major mineral and energy resources, albeit in a somewhat erratic distribution. There are also rich forests and fertile soils but there is little point in mapping these attributes as their exploitation has depended very much on the principal resource of the region – its people. It is their inventiveness and energy over the years that have transformed riverine flood plains or seasonally dry regions into fertile and productive areas through irrigation and drainage, and it is their labour that currently creates profit where no physical resources exist.

Pacific Asia is also important in terms of physical resource output, although it must be admitted that it is countries that do not lie within the purview of this book, such as China, Australia and the Soviet Far East, from which most of the regional output derives. Nevertheless, major oil resources are currently being exploited and exported in Brunei, Indonesia and Malaysia, whilst several countries feature in the list of top ten producers of nickel (Indonesia, Thailand) and tin (Malaysia and Thailand).

Agricultural exports are also substantial, with the region dominating world production of a number of key industrial commodities such as rubber, palm oil and copra, as well as continuing to produce its traditional export crops of rice, tropical fruits and various spices. One problem with industrial agricultural resources has been the fact that prices have fallen dramatically over the last few decades. Far from discouraging production, this has pushed it to ever-higher levels in order to ensure that revenues are retained (Figure 1.5) – often at the expense of other environments (see Chapter 3).

Of course it can be fairly claimed that the intensity of development of both agriculture (and industry) has been the consequence of sheer population pressure on resources. Although death rates began to fall in the late colonial period, birth rates stayed high almost everywhere until the 1970s. Even so, the rate of natural increase is still high in many countries and has posed serious problems for the governments of Indonesia, the Philippines and other states (see Chapter 5).

The nature of the human resources of a region are, however, dependent upon much more than sheer numbers. The quality of life and the labour potential is inextricably linked to the levels of education, nutrition and health care (Table 1.1), all of which vary enormously, not necessarily correlating to the level of development (as indicated by GNP). The nature of this variation is discussed below. There are, however, less quantifiable aspects to the human and physical resources of the region. These might relate, for example, to the cultural mix, the architectural legacies of the past and the beautiful scenery. The variety of each encompassed in Pacific Asia is probably far more extensive than in any other part of the developing world – a feature indicated by the massive growth in tourism. On the other hand, both the human and physical resources in the region have a certain fragility, due in part to their complexity and symbiotic relationship with one another. One consequence, therefore, of the rapid changes of the post-independence years has been both environmental and human exploitation, subjects which are investigated in Chapter 3.

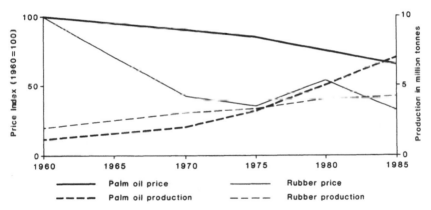

Figure 1.5 Palm oil and rubber: price and production trends 1960–85

The political situation

In many ways the variations in the development process in Pacific Asia reflect its political composition. It is perhaps an oversimplification merely to divide the region into its communist and capitalist components. The socialist states are all very different in their nature, with China and Vietnam constituting traditionally bitter enemies; North Korea pursuing a dated but not unsuccessful Stalinist development strategy; and Burma blending socialism and Buddhism into a unique brand of isolationist stagnation. Nor can it be said that the capitalist states are any more

Table 1.1 Pacific Asia: selected indicators of growth and development

	Demographic					GNP			Economic	GDP%			
	Pop. (m)	CBR/ '000	CDR/ '000	% pop under 15	% urban	pc US$	Av. ann. % growth 1970s	1980s	as % of OECD av.	Primary	Ind.	(Mfg)	Tertiary
Brunei	0.2	30	4	38	64	15,500	–	–4.8	106	–	–	–	–
Burma	38.8	34	13	39	24	190	1.2	5.5	1	48	13	–	39
Cambodia	6.5	39	18	35	11	–	–	–	–	–	–	–	–
China	162.0	21	8	28	32	290	–	10.4	2	31	49	34	20
Hong Kong	5.6	14	5	24	92	8,070	6.8	6.6	55	0	29	22	70
Indonesia	174.9	31	10	40	22	450	4.0	3.5	3	26	33	14	41
N. Korea	21.1	30	5	39	64	1,123	–	2.0	8	–	–	–	–
S. Korea	42.1	20	6	31	65	2,690	7.0	8.8	18	11	43	30	46
Laos	3.8	41	16	43	16	160	–	7.0	1	63	6	–	31
Macao	0.4	23	6	34	97	–	–	8.0	–	0	35	–	65
Malaysia	16.1	31	7	39	32	1,810	4.3	3.2	12	21	21	–	58
PNG	3.6	36	12	42	13	700	2.8	–	5	34	26	9	40
Philippines	61.5	35	7	41	40	590	2.8	–1.1	4	24	33	25	43
Singapore	2.6	17	5	24	100	7,940	7.5	7.2	54	1	38	29	62
Taiwan	20.3	17	5	30	67	3,470	8.0	7.5	24	6	51	32	43
Thailand	53.6	29	8	36	17	850	4.7	4.1	6	16	35	24	49
Vietnam	62.2	34	8	40	19	0	–	–	4	45	26	–	29
Pacific Asia		23	7	–	37	470	–	7.0	3	21	45	–	35
Japan	122.2	12	6	22	76	15,760	7.1	3.7	107	3	41	29	57
Third World		30	10	–	37	700	–	4.0	5	–	–	18	–

Notes: Industry includes manufacturing (mfg) and mining; primary includes agriculture, forestry & fishing

Social

	Life exp.	IMR/ '000	Daily cal. intake	Persons per doctor	% primary		% secondary		% share h'hold inc bottom 40%	top 10%
					M	F	M	F		
Brunei	74	12	–	2,000	–	–	–	–	–	–
Burma	53	103	2,600	3,700	–	–	–	–	–	–
Cambodia	43	160	–	–	–	–	–	–	–	–
China	66	61	2,600	700	137	120	48	35	–	–
Hong Kong	75	8	2,900	1,100	106	134	66	72	16	31
Indonesia	58	88	2,600	–	121	116	45	34	14	34
N. Korea	69	33	3,200	420	–	–	–	–	–	–
S. Korea	67	30	2,900	1,000	94	94	98	93	17	28
Laos	50	122	–	5,000	–	–	–	–	–	–
Macao	68	12	–	2,300	–	–	–	–	–	–
Malaysia	67	30	2,700	3,000	100	99	54	54	11	40
PNG	55	100	2,200	1,000	–	–	–	–	–	–
Philippines	65	50	2,300	6,700	107	106	66	69	14	37
Singapore	71	9	2,800	1,000	118	113	70	73	–	–
Taiwan	73	9	–	1,100	–	–	–	–	–	–
Thailand	63	57	2,300	6,000		99		29	15	34
Vietnam	63	55	2,300	3,300	107	94	44	41	–	–
Pacific Asia	68	–	2,600	2,400	131	117	50	39	–	–
Japan	78	6	2,900	670	101	102	95	97	22	22
Third World	62	–	2,509	4,630	112	94	47	34	–	–

Notes: Education more than 100% is recorded when people outside normal age group are in that section.
– denotes data unavailable

Sources: Far Eastern Economic Review Yearbook 1990
World Development Report 1990

unified. The principal internal alliance is ASEAN (Association of South East Asian Nations) but it has taken a long time to overcome the traditional antipathy and suspicion between these Southeast Asian neighbours.

Within the capitalist camp too, other events have created tensions. For 40 years the United States has been the principal military backer of the capitalist states; indeed, it has not only bankrolled the defence of the region but has also funded economic development of the most strategically important states, such as South Korea and Thailand. At present, the United States sees itself being displaced by Japan as the dominant economic influence in the region and fears that the growing economic prosperity of other states may lead to assertions of military independence too. In particular, the United States is worried about the future of its huge naval/air base at Clark Field in the Philippines. Facing this across the South China Sea is an equally large Soviet base at Cam Ranh Bay in Vietnam (ironically once a major American installation).

But the posturing of the superpowers in Southeast Asia is only one of the many political conflicts of the region. Although to investors it may appear relatively stable compared to other parts of the Third World, this is far from the case. Only Malaysia, of the principal countries in Pacific Asia, has never had direct military intervention in government. Elsewhere military regimes, often brutal and repressive, have been all too common.

Potential trouble spots exist aplenty (Figure 1.6). Almost all capitalist states have experienced or are continuing to experience communist-inspired insurgency movements whilst, in their turn, many communist states are subject to pressures for democratic reforms which have received at best unsympathetic responses, and at worst brutal suppression. Many specific unresolved political issues remain to pose potential problems for the future – Hong Kong, Macao, Taiwan and a divided Korea. Little wonder that the levels of military expenditure in the region are very high and for many countries constitute a severe impediment to increased economic growth.

Levels of development and standards of living

In seeking to compare and evaluate development, it must be remembered that the available statistics are very rarely collected according to a consistent set of standards and definitions. Each country adopts its own set of measurements, some of which may be compatible with those of other countries, some of which may not. For example, almost every

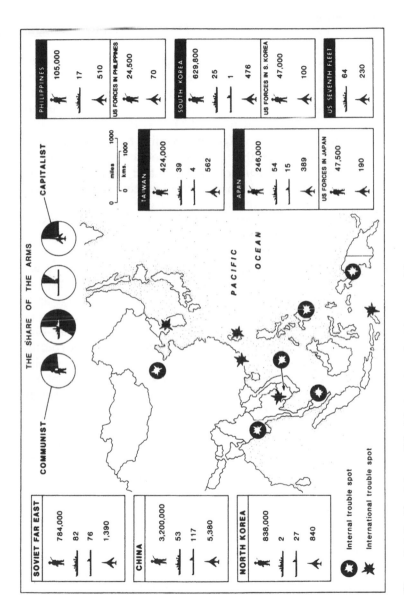

Figure 1.6 Pacific Asia: political hot-spots and major armed forces
Source: Far Eastern Economic Review

country differs as to what constitutes an 'urban' (as opposed to rural) population; whereas infant mortality rates are more likely to be compatible in definition, if not in collection.

The statistics collated in Table 1.1 are subject to all these criticisms but provide a useful basis on which to construct a generalized overview of development in Pacific Asia. Large gaps exist in the data, particularly for the socialist states who do not make their information (if collected) available to the western sources from which this table was compiled. Most of the aggregated data are weighted according to the population of individual states – this means that overall regional figures are heavily skewed towards the characteristics of Indonesia and, more particularly, China. It is also unfortunate that the World Bank continues to pretend that Taiwan does not exist as a separate state, largely because it does not wish to offend China. This is a particular difficulty in analysing industrial growth and the reader ought to bear in mind that data on Taiwan may come from a variety of less formal sources.

On examining Table 1.1 it is immediately apparent that there is a substantial difference in per capita GNP between the socialist and capitalist states. In the past, some would have interpreted this as an indication that the socialist states are still very backward and underdeveloped. The composition of GDP might seem to bear this out, since the highest contribution from agriculture can be observed in the socialist countries. However, this is not necessarily representative of underdevelopment so much as different priorities. A glance at the social data will reveal that the distinction between the two political groups is far less clear cut. The countries least provided with health care or educational facilities are drawn from both. Indeed, some socialist states appear to be better-off in terms of social indicators than even the wealthiest capitalist states. Is this the result of deliberate government policy or do some states experience more difficulty than others in providing a better social infrastructure?

You might now expect, given these observations, that the capitalist nations in Pacific Asia do not exhibit much uniformity in their development indicators, and you would be correct. Table 1.1 reveals considerable diversity in the nature of economic growth, not only in levels of GNP but also in its components in both respects. Hong Kong and Singapore stand out as being particularly prosperous and virtually devoid of any primary sector (as one might expect in city-states). It is worth noting in this respect that their economies are not dominated by manufacturing, as popular opinion might suggest, but by their tertiary or service sectors. However, this sector is not uniform in its coverage, encompassing traditional retailing activities, entertainment and transport services, as well as office activities. It is this last component which makes Hong Kong and Singapore so affluent since it reflects the influx of international banks and headquarters of multinational firms that service regional economic growth – producer services as they are usually known. What is *not* included in the GDP data is the contribution made by what is commonly termed the informal sector. These comprise non-wage activities, often illegal, which enable urban residents to make some sort of living, and result from the rate of population growth being much more rapid than the rate of job creation.

At this point it is as well to inject a note of caution in interpreting national resource data. Although the region's average per capita GNP seems far below the Third World average, this is weighted considerably by the figures for China and Indonesia. Individually many states in the region have much higher incomes and are classed by the World Bank as middle or upper-middle income. Indeed, in the 1989 edition, Hong Kong and Singapore were for the first time been grouped with the high-income economies of the world. And yet even these high-flyers, with their sophisticated economies and affluent lifestyle, attain only just over half the average per capita GNP of OECD (the advanced capitalist) nations.

Nevertheless, the extent to which all of the nations of Pacific Asia are tied to the world economy can be seen in recent trends. In all the states, agriculture and minerals have declined as a contributor to GDP, partly owing to the redirection of investment into industry, but also because of the steady decline in world commodity prices; this is very evident in the oil-based economy of Brunei. However, the global depression of the 1980s has affected all capitalist economies and, compared to the 1970s, economic growth rates are down, particularly where political instability has further frightened away investors, as it has in the Philippines.

The slow rate of economic development in some parts of Pacific Asia is clearly the result of population pressure on resources. Overall birth rates in the region are amongst the lowest in the Third World, a reflection of both levels of economic growth and the introduction of fertility control programmes, but so are death rates, due to the high priority given to improving health care and education. Indeed, in Singapore and Malaysia the fall in birth rates has been such that their governments have put fertility control policies into reverse, fearing shortfalls in their future labour force (see Chapter 5). However, many other countries still have very youthful populations, with around 40 per cent under 15 years of age – a new and extensive generation of potential parents already exists.

In addition to uneven development within Pacific Asia, there are also disparities within individual countries. Some of this is spatial or geographical unevenness, with core regions around large cities receiving most of the national investment. Another type of disparity is more social in origin. As the final columns in Table 1.1 clearly indicate (for those states for which data are available), growth and equality do not necessarily go hand in hand. Much of this inequality is not simply class based; it reflects other divisions in society too. Ethnicity is one of these (see Chapter 6), so is gender (see Chapter 9). The outcome can be an extremely complex relationship between the social and economic aspects of development. It is to the evolution of this complex relationship that we now turn.

Key ideas

1 Pacific Asia has strong trading links with developed countries in both primary and manufactured exports.
2 The principal resource of the region is its population.
3 The political situation is still not stable.
4 Standards of living in Pacific Asia are rising but still vary considerably throughout the region and within individual countries.

2
An historical geography of Pacific Asia

Introduction: definitions and framework of analysis

Any discussion of the historical geography of the Third World is usually liberally sprinkled with words such as 'imperialism', 'colonialism', 'neo-colonialism' and so on. In some texts it is almost impossible to distinguish a separate meaning for each and, indeed, the terms will often have been used interchangeably. But what do they mean and how do they differ?

Imperialism

Imperialism tends to be used in two different ways. First, there is the general or colloquial sense which describes the exploitative relationship between the core and periphery. Second, in a technical sense as used by Marxists, to discuss the latest stage in the evolution of capitalism – a stage in which a high degree of concentration of the ownership of the means of production has resulted in monopoly capitalism, or control of the world economy.

Clearly these two alternatives are not incompatible but there is considerable disagreement over the dates to which it applies, even amongst Marxists. Some claim that imperialism began only towards the end of the nineteenth century, others that it emerged when capitalism started to expand out of Europe in the sixteenth century, following the demise of feudalism.

Colonialism

Colonialism is the establishment of control by one society over another. This can occur for economic reasons (one of the prime incentives within imperialist expansion), but it can also result from purely political motives too: for example, for strategic purposes, for defence of frontiers, as a diplomatic manoeuvre and so on. Colonialism is, therefore, much older than imperialism both as a concept and as an historical process.

Colonialism became much more widespread in the nineteenth century in association with the rapid expansion of capitalism. It is this fusion of political and economic forces which has given rise not only to the interchangeability of the terms colonialism and imperialism, but also to the incorrect assumption that colonialism began in the nineteenth century.

Framework of analysis

Clearly there is considerable overlap between the different types and phases of both colonialism and imperialism. How can this be resolved into a simple and useful framework for investigation? One problem is clearly posed by the fact that the major regions of the Third World experienced imperialism and colonialism over different chronological periods: Latin America well before Asia, and Africa last of all. The model suggested below, therefore, is primarily structured around the time-scale for Asia, although this does not invalidate its use elsewhere.

Figure 2.1 illustrates the way in which the types and phases of both imperialism and colonialism have been incorporated into a chronological model of the development process in Pacific Asia from the beginning of the sixteenth century when the first major commercial and trading contacts were made. It is important to appreciate fully that substantial, sophisticated and complex civilizations existed in Pacific Asia prior to the arrival of the Europeans and that the reaction of these societies to European demands for trade varied enormously.

In general, the long period from 1500 to 1800 is marked by the slow expansion of European capitalist interests in Pacific Asia. The late eighteenth and early nineteenth centuries witnessed a transition in both imperialism and colonialism but by the mid-nineteenth century the demands of industrialization in Europe had led to a scramble for territory rather than trade. Imperialism became dominant and colonialism became formal and direct as the European powers

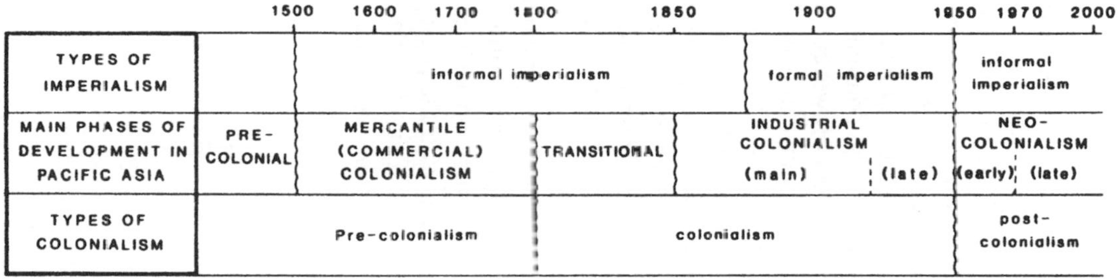

	1500	1600	1700	1800	1850	1900	1950	1970	2000
TYPES OF IMPERIALISM	informal imperialism					formal imperialism	informal imperialism		
MAIN PHASES OF DEVELOPMENT IN PACIFIC ASIA	PRE-COLONIAL	MERCANTILE (COMMERCIAL) COLONIALISM		TRANSITIONAL		INDUSTRIAL COLONIALISM (main) (late)	NEO-COLONIALISM (early); (late)		
TYPES OF COLONIALISM		Pre-colonialism		colonialism			post-colonialism		

Figure 2.1 Colonialism: a framework for analysis

(later joined by the US and Japan) sought to acquire large slices of Pacific Asia in order to exploit their resource potential. What is perhaps worth noting at this stage is the relatively brief chronological span of industrial colonialism. Indeed, its main phase, during which colonies were established and economies restructured, was even shorter, giving way after the shattering effects of the First World War to a much different, later phase of administrative consolidation.

Although the Second World War effectively brought an end to direct colonial control, it did not and could not halt the expansion of capitalism. Despite independence, therefore, the economies of most countries in Pacific Asia continue to be dominated by enterprises whose headquarters are located overseas. However, even in this relatively short period of 40 years there has been a noticeable shift in the nature of neo-colonial exploitation. For the first 20 years or so it remained similar to that of the first half of the century, being structured around the acquisition of primary commodities for processing in the advanced nations. Since 1970, in contrast, much of the manufacturing process has moved to the Third World itself, in search of cheap labour. This process has been selective but nowhere has it had more impact than in Pacific Asia.

These, then, are the main phases that we will use both to describe the evolution of contemporary Pacific Asia, in order to make sense of the legacy of the past, and to structure analysis and explanation in the remainder of the book. We begin with the pre-colonial period in order to remind ourselves that there was considerable life in the region prior to the arrival of Europeans and their economic system.

The pre-colonial period

European colonialism in Pacific Asia began in the early sixteenth century so that the pre-colonial era relates to a much earlier period than is commonly supposed. This is not to deny European contact prior to 1500, but from the sixteenth century onwards such contacts were firmly part of a process of capitalist expansion.

Clearly, Pacific Asian civilizations prior to this date varied enormously over space and through time. It was certainly dominated by the Chinese empire which considered itself to be the centre of the universe and totally self-contained. Even in 1793, following almost three centuries of European trading in Asia, the Emperor Ch'en Lung could inform George III (through his emissary Earl Macartney) that:

Our celestial empire possesses all things in prolific abundance and lacks no product within its borders. There is therefore no need to import the manufactures of outside barbarians in exchange for our own produce. But as tea, silk and porcelain . . . are absolute necessities to European nations, and yourselves, we have permitted . . . your wants [to] be supplied and your country [to] participate in our beneficence.

(quoted in Osborne, 1985)

However, elsewhere substantial civilizations and empires developed that were equal to, and often surpassed, anything that had emerged in Europe since the Roman period. Thus, as Milton Osborne remarks in his *Illustrated History of Southeast Asia*, in 1066 when William the Conqueror was being crowned in a shambling London of some 35,000 souls, more than a million lived in Angkor, the flourishing capital of the Khmers (Case study A). Elsewhere and at other times, similarly grand capitals existed, such as Pagan in Burma, easily comparable to the best of Europe's own classical periods.

Case study A

Angkor, Cambodia: a pre-colonial empire

The Angkorian empire lasted for some 600 years, from the ninth to the fifteenth century. When the first Europeans arrived they saw only the impressive temple that had once been the centre of an extensive city and kingdom in which culture and technology were developed to the highest levels.

The key to the achievements of the Khmer people was their refinement of what has been called an 'hydraulic society': one in which the seasonal rainfall was harnessed by a series of dams, reservoirs and irrigation canals to increase dramatically the rice production of the fertile Cambodian plains around the huge lake of Tonle Sap. The irrigation technology had been developed gradually over many centuries, and the Cambodian area had witnessed the rise of a series of civilizations, such as Funan, Chenla and Champa. Angkor was to be the principal empire that arose in this part of Southeast Asia, until the migrant Thais began to settle and challenge it. At its greatest extent, almost 170,000 hectares were under irrigation along the northern side of the Tonle Sap alone.

Case study A *(continued)*

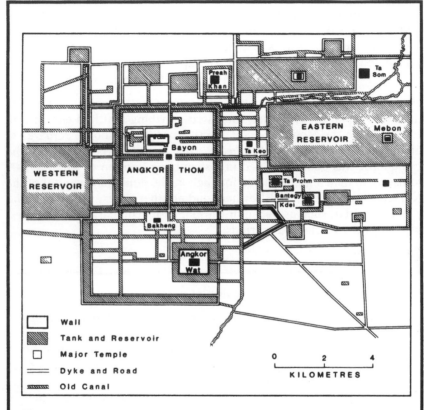

Figure A.1 Angkor: city plan
Source: T. G. McGee *Southeast Asian City*, Bell, 1967

The food produced in this way enabled some people to leave agricultural labour and sustained a whole series of non-subsistence occupations, such as priests, soldiers, craftsmen, etc. Angkor was not the centre of a trading empire, it was a theocracy with religion and its associated beliefs and obligations binding together the entire society, the physical focus of which was the temple complex in Angkor (Figure A.1).

At its height in the twelfth century, Angkor was surrounded by

Case study A *(continued)*

vassal states who recognized the superiority of its ruler and sent tributes. However, once the hydraulic technology of expanded food production spread to other fertile areas within the region, this relationship waned and with it the wealth and power of Angkor. Indeed, once the irrigation systems themselves were disrupted, not only the empire disappeared but also much of the city of Angkor itself. Only the extensive temple complexes remain, although the early Europeans were not much impressed since there were few commodities to trade. Nevertheless, at its height, Angkor must have been far more impressive than any twelfth-century European state.

Apart from China, much of this historical splendour was not evident when the first regular European contacts with the region began in the sixteenth century, as various events had combined to produce a much more fragmented mosaic of states. Forces acting towards both unity and disunity exist within any region and Pacific Asia is no exception. In general terms, the region was dominated by Indian and Chinese cultural and political influences. Those of the Chinese empire were surprisingly limited outside China itself (largely to Vietnam and Korea), presumably as a consequence of its wealth, which reduced expansionist urges.

Elsewhere in Southeast Asia, Indian influences were predominant, not only in terms of the spread of Buddhism and Hinduism but also in terms of the political system which accompanied them. As a result, most Southeast Asian states were structured around a hierarchical system headed by a (semi-divine) king, followed by an aristocracy, then by a level of priests, army and civil service, all superimposed on a massive peasantry. The whole system was also spatially centralized around a primate settlement which functioned both as an elite capital and as a ceremonial/religious centre (Plate 2.1).

By the time the European traders arrived at the beginning of the sixteenth century, the situation in parts of Pacific Asia was much more fragmented, particularly in Southeast Asia where no powerful nations remained to impress the early arrivals whose accounts have coloured

Plate 2.1 Borobodur: the pre-colonial temple complex. The scale and complexity of this temple clearly indicate the sophistication of this Buddhist state of which it was the focus.

present-day interpretations of the contemporary scene. In many ways such impressions were reinforced by both conservative and radical/Marxist development strategists who considered Asiatic society to be stagnant, static and feudal in character (see Dixon 1991).

These impressions were not entirely accurate. True, the Burmese or Khmer empires were not what they once were, but the sixteenth-century rulers were very much aware of their inherited status. Much of the political change had been consequent on Mongol penetration into Northern Burma; in turn Burmese and Thais had pushed into other mainland core areas. Moreover a new democratic form of Buddhism began to penetrate the region, eroding central state controls and administration. Finally, during the fourteenth and fifteenth centuries the wealth and commercial success of Arab and Indian traders resulted in the rapid spread of Islam throughout the Southeast Asian archipelago

Although such events gave rise to political instability, this was no different to the situation in contemporary Europe. It was into this general state of political flux that the Europeans intruded.

Mercantile colonialism

Feudal Europe had set the stage for mercantile colonialism with the development of the market economy based on the city and the subsequent rise in status of commercial and merchant classes. The technological revolution at the end of the fifteenth century signalled the demise of the city-state but ensured that the commercial capitals remained the focus for the aggressive new European nation states which began to encourage their merchants to seek wider spheres of activity.

The initial moves overseas were made to acquire either precious metals (gold and silver) or the more valuable items of the European trading system, such as porcelain, silks and above all spices. This took merchants towards Pacific Asia, and particularly towards Southeast Asia where the spices were produced and where Chinese goods were available from other regional traders. Immense profits were to be made out of this trade and, given the demand for spices, the difficulties of the trading system and the large number of individuals involved, it is not surprising that prices and profits in Europe were very high; pepper was the most speculative commodity of the sixteenth century.

Early trading activities were dominated by the Portuguese who, in the hundred years following Vasco da Gama's voyages, dominated direct European links with Pacific Asia to the extent that Lisbon was known as the spice capital of Europe. Portugal was in fact involved in much of the other trade within the region, for example between India and China, and between China and Japan. But in all of this activity Portuguese traders comprised only a limited proportion and maintained their domination of the European trade through a chain of small port concessions which were sustained by playing off one local state against another.

Spanish attempts to take over Portuguese trading activity were strongly rebuffed and eventually Spain settled for establishing a regional base in the Philippines, one of the least resourced and most fragmented parts of the region. Here in Manila Spain set up a collecting point for regional produce which was then shipped once a year to Acapulco in Mexico (or New Spain as it then was), from where the Philippines was administered. In contrast to other European powers, there was more settlement in Manila, which became a city of 40,000 by 1620. Significantly, there was a substantial religious presence in the Philippines, currently indicated by the fact that it is still the only predominantly Christian country in the region.

By the seventeenth century, changes in the European power structure were beginning to filter through to Asia, and the leading European traders in the region became the Dutch. Dutch trade was much more organized than that of the Iberian nations, both in Europe and in Pacific Asia, and the seventeenth century witnessed the first structured expansion of European capitalism into this part of the world. For the most part, this was accomplished under an East India Company, as was later English rivalry, which operated a systematic war of attrition against both the Portuguese and the local rulers whom, in contrast to earlier years they sought to subdue by force of arms.

The result of this intensification of activity was not only increased profits but also increased commitment to the assembly, storage and protection of commodities. The escalating cost of administrative and military functions began to weigh heavily on the company, particularly when English incursions began to threaten both trading monopolies and bases. In addition, the incentive to move into actual production in order to increase both the quantity and quality of the goods sold also began to push operating costs way beyond the companies' means: by the beginning of the nineteenth century both the Dutch and English East India Companies were virtually bankrupt.

In this discussion of expanded European activities we must not overemphasize their physical presence. They were still limited to small sections of a few cities, apart from Manila, in the islands and peninsulas of Southeast Asia; very little direct impact was felt on the mainland from Burma through to Korea. Moreover, most of the extensive regional trade was still in the hands of non-Europeans and many of the rulers felt no real threat from European activities.

Yet change had occurred. Trading networks were now larger than ever and had imperceptibly been drawn increasingly into the demands of European capitalism. Fragmentation of states had occurred as a result of Portuguese, Dutch and English pressure for allowances and concessions; the nature of warfare had escalated by the use of European weaponry, tactics and personnel; and new ports had emerged in response to the changing political and economic geography of the region. Even in areas which had not directly felt the impact of

Europeans, new pragmatic rulers were emerging as a result of the need to be aware of the growing accumulation of economic and political pressures. Thus in Thailand in the mid-eighteenth century the present Chakkri dynasty was established through a series of reformist monarchs. Even in China, the Emperor had by now realized that Europeans would not die of constipation if deprived of rhubarb!

As yet, however, these changes in thought and trade impinged directly on only a relatively small number of people in Pacific Asia. For the great mass of peasants and their local masters, life seemed to continue very much as it had done for thousands of years. The difference was that their activities, whether subsistence or market oriented, had over the long period of mercantile colonialism (some 300 years) been subtly linked to a nascent world economy, the core of which lay in Europe.

But for some time in the nineteenth century, it appeared as if further change in Asia would be quite limited. The costs of organizing colonialism had moved beyond the means of private companies; whilst Europe was involved in a massive continental war which consumed state resources and attracted the individual adventurers on which so much mercantile activity had depended. But, above all, the European industrial revolution offered more lucrative opportunities for investment. Indeed, most of the accumulated mercantile capital was used to finance the acceleration of industrial production. It was not until the middle of the nineteenth century that changes within Europe itself caused investors and governments to look afresh at Pacific Asia and other parts of the world.

Industrial colonialism

The period which is commonly recognized as the epitome of explorative colonialism in Asia and Africa was quite short-lived. For some, like Lenin, it did not begin until the Congress of Berlin and partition of Africa in 1885, and by 1920 the era of continuous profitability was largely over. Most observers, however, date the commencement of the industrial or high colonial era to the mid-nineteenth century when European, primarily British, industrial capital began to look for ways of expanding production to retain profitability. Two obvious ways were to seek cheaper raw materials and new markets overseas. In addition, if cheaper food could be obtained to feed the growing urban workforce, then the costs of reproducing labour could also be kept low.

The sources of these cheap raw materials, food and, eventually, extended markets were to be in restructured colonies. In place of the company, there would be the national state. For a while Britain hoped to tap the resources of Asia and Africa by extended 'free trade' but by 1875 it became clear that, given the rise of rival industrial states within Europe, Asian and African resources would have to be more firmly attached to the metropolitan power. This was done by territorial occupation. Prior to 1870, annexation and occupation tended to follow resource exploitation; after this date it tended to precede it.

Territorial occupation occurred for a wide variety of reasons. Many

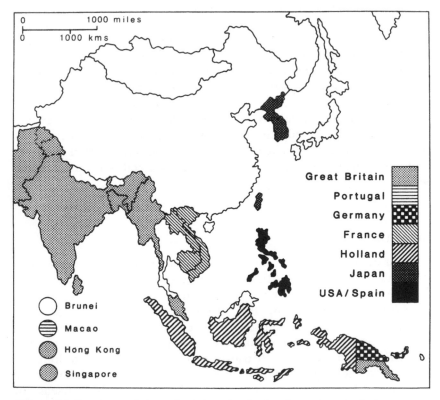

Figure 2.2 Southeast Asia: colonial territories by the early twentieth century

were, indeed, economic, such as the desire to exploit mineral resources or reorganize agriculture production to produce a commercially exportable crop. However, other reasons became equally important once the scramble for territory accelerated – the strategic desire to maintain lines of communication or protect existing colonies (or to deny others the prospect of these) or simply to satisfy national pride in acquiring one's 'rightful share' of colonies.

Within Pacific Asia the process of accelerated territorial acquisition followed the opening of the Suez Canal in 1869 and the establishment of telegraphic links. Very quickly the major industrial powers carved out for themselves enclaves of influence (Figure 2.2) into which remnants of earlier Iberian and Dutch colonialism intruded. The principal areas left for territorial acquisition lay on the Asian mainland where the main prize was China itself. Despite its many problems and weaknesses, China managed to hold on to its territorial integrity, although it was forced to concede a series of treaty ports to the major powers. Britain and France, therefore, sought other avenues into the resources of the interior of China, notably through the major river valley systems of Southeast Asia. As Britain consolidated her hold in the western part of this region, so France re-created her Asian colonial interests in Vietnam (see Case study B). Meanwhile, the Dutch had expanded from their Javanese base to annexe a large number of the islands off the Southeast Asian coast. Germany attempted a late and short-lived incursion into New Guinea but the other major European power, Russia, found herself thwarted by the rapid rise of a new aggressive colonial state in Japan which rapidly carved out its own territorial identity in Korea and Formosa (Taiwan). Yet another Pacific colonial power arrived in the region shortly before the end of the century when the United States purchased the Philippines (then in the throes of revolution) from a bankrupt Spain.

Case study B

Vietnam: colonialism and its impact

When the French arrived in Vietnam in the nineteenth century, they found a long-established and fiercely independent state which, although it was strongly influenced by Chinese ideas and

Case study B *(continued)*

was, indeed, a tributary state of the Celestial Empire, nevertheless had succeeded in maintaining its independence since AD 939. Unlike other parts of Southeast Asia where Indian influences were paramount, Vietnam was held together by a vast bureaucracy, entered on merit by examination, which linked every aspect of the state through Confucian principles to a ruler known as the Son of Heaven.

France invaded Vietnam in 1859, partly to thwart British colonial advances in Southeast Asia and partly to seek a route into China, little realizing the traditional antipathy between the two countries (which still persists today). At the time of France's invasion, Vietnam was an expanding but conservative empire with its capital at Hue in the centre of the country. Unlike most other colonial nations, France was invading a well-established political entity.

Shortly after capturing Saigon, France established her colony of Cochin China, together with a protectorate over Cambodia. With the Mekong route into China proving less than satisfactory, the French sought an alternative through the Song-koi or Red River in the north and bombarded Hue until the two provinces of Tonkin and Annam were ceded (Figure B.1). In 1887 all three were combined into the Indo-Chinese Union to which was later added Laos. Within just 25 years France had complete control over a well-organized and locally powerful state whose conservative political ideology had proven incapable of providing any real resistance.

As Bill Kirk (1990) has drily observed 'of all the colonial powers in Southeast Asia ... France was the most hierarchical administratively, exploitative economically and elitist culturally.' The Vietnamese simply had no say in what happened in their own country, building up a considerable resentment which effectively exploded in the post-war years.

French development of Vietnam was regionally quite different. The sparsely populated Mekong delta was drastically changed through new hydraulic engineering and the consequent granting of huge estates to both French and Vietnamese landlords. These were farmed by tenants who paid 40 per cent of their crop in rent.

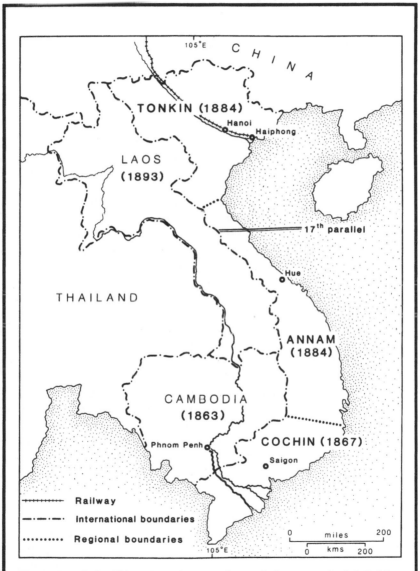

Figure B.1 Indo-China: French expansion and the post-colonial division of Vietnam

Case study B *(continued)*

The infrastructural developments also led to the introduction of plantation agriculture, primarily coffee, tea and rubber, which used contracted labour from north Vietnam. By 1938 the delta produced three-quarters of French Indo-China's exports by value, of which half were rice, with most of the remainder constituted by rubber and maize. Towns in the south were relatively small and dominated by ethnic Chinese.

After the capture of the north, its greater population and mineral resources began to attract increasing French investment and in 1902 the capital was moved to Hanoi, which was virtually rebuilt as a French city. The capital and associated centres, such as the port of Haiphong, became the industrial and commercial heart of the Union but, although they became a magnet for migration, rural population pressure built up enormously as both diseases and natural hazards were brought under control. By 1939 the Red River delta had more than 6.5 million people. As a result of the consequent subdivision of holdings, by 1940 more than 90 per cent of the northern peasants farmed less than one acre. Although poverty, debt and landlessness became common, the basic social unit of the commune remained substantially intact, and it was in the north rather than the south that Vietnamese national identity remained strong.

During the Second World War, the Japanese allowed the French to retain nominal control in Vietnam but assumed power in 1945 through the medium of declaring Vietnam to be independent under Emperor Bao Dai. Six months later the Japanese surrendered and the Viet Minh, still predominantly nationalist but led by the Ho Chi Minh, seized control. The French returned soon afterwards and assisted initially by British and Nationalist Chinese forces, proceeded to try to reclaim its Union. The Viet Minh fought a guerrilla war for eight years until their growing strength and increasing French disillusion led to a major military defeat for the colonialists. In 1954 the north became independent whilst a US-backed government was set up south of the 17th parallel (Figure B.1). It was to be another 21 years of dreadful conflict before the whole country was finally united.

Despite the haste of the scramble for territory, the major European powers were able to retain some buffer zones in between their principal spheres of interest in order to minimize what could prove to be costly disputes. Thus, Thailand was never colonized, with Britain and France acting on several occasions as guarantors for its boundaries (against incursions by Belgium, as well as by another), whilst North Borneo was left between the Dutch East Indies, British Malaya and the Spanish Philippines. But whichever European nation was involved, the political geography of the region was completely recast with little reference to previous physical, ethnic or cultural factors – with enormous repercussions for the contemporary situation.

Given the variety of colonial powers and the differences in the nature of the indigenous states with which they came into contact and conflict, industrial colonialism in Pacific Asia was a complex and varied process. As noted above, the British preferred indirect rule, at least at first, whereas France and Holland imposed a much more direct, centralized metropolitan domination over their colonies. Even within the same metropolitan sphere of influence and activity, the nature of colonialism varied over time and across space. In some locations, agricultural restructuring occurred through the creation of large-scale plantations. In others, local producers were encouraged to amalgamate holdings; elsewhere large-scale European settlement occurred (although not to the same extent as in Algeria or Southern Africa).

Clearly the nature of the system that was adopted resulted in important differences in the impact on local populations. Particularly traumatic were the demands for labour that were generated by mining and plantation agriculture. This was undertaken either by imposing tax demands, which forced indigenous farmers into wage labour in order to raise the necessary cash, or by wholesale restructuring of agricultural tenure to create a class of landless labourers, or by importing labour from outside the region or even outside the country.

In terms of production and exports, therefore, many colonies witnessed a rapid narrowing in the range of commodities available. Colonies became associated with the production of one or two items and imported whatever else they needed. Needless to say, metropolitan firms were in control of import–export activities and reaped most of the profits.

It would be a mistake to think that this transformation involved only new products. In some colonies rubber, oil palm or coffee were the leading export-earners but in others it was more traditional crops, such as coconuts, which satisfied the increasing demand for (in this case) the raw materials for soap manufacture. Even rice became an important export-earner as immigrant labour groups and accelerating urbanization increased demands within Asia itself. This occurred in Thailand (Figure 2.3), despite the fact that it was not a colony, with expatriate firms, particularly British, dominating the export of rice to other destinations in Pacific Asia where it was used, in effect, to lower the costs of colonial exploitation.

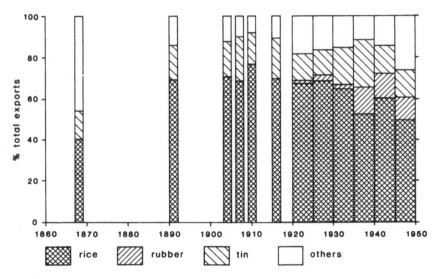

Figure 2.3 Thailand: rice and commodity exports
Source: Dixon 1991

As a result of the drastic economic, social and demographic changes of industrial colonialism, the last quarter of the nineteenth century also witnessed the acceleration of market potential in Pacific Asia for western manufactured products. Indeed, it was the purchasing power of the Chinese and Japanese markets that was as important as access to their products in encouraging the western powers in their almost frantic attempts to gain trading concessions. In the colonies themselves, the initial markets for western goods were confined to wealthy expatriate and indigenous elites. But the quality and price of these goods and the demonstration effect of purchases by the wealthy soon resulted in imported commodities dominating the expenditure pattern of all social groups, even rural subsistence, thus further destroying the indigenous artisan economy and leading to an increasing dependency on the west. Particularly poignant in these circumstances was the re-export of cheap food to the growing markets amongst the urban and rural poor. Their diet of flour, sugar and tea often had colonial origins but was processed (and value added) in Europe, thus facilitating a double exploitation of the colonial poor (through their labour in growing the crops, and through their subsequent purchase of it at exorbitant prices).

The corollary of this situation is that manufacturing was relatively limited during the industrial colonial phase. Any manufacturing that existed was largely concerned with the preliminary processing of primary products, such as rice milling or tin smelting. But most of the more sophisticated processing (and creation of profits) occurred within the larger parts of the metropolitan country. This is the era during which the big dockside manufacturing plants for tobacco, sugar, etc. proliferated in Liverpool, Glasgow and London.

However, it would not be correct to assume that colonial cities were simply points of control and administration. Although few were centres of production, commercial activity, from the manufacture of small consumer goods to the retailing of imported products, was very extensive. Much of this activity was, of course, in the hands of non-Europeans. This is not the same as saying they were the prerogative of local entrepreneurs, because almost all of the colonial powers in Pacific Asia, other than the Japanese, made a point of encouraging or permitting immigrant groups, usually Chinese or Indian, to infiltrate and monopolize local commerce. In this way, a convenient demographic, cultural and economic buffer was placed between the colonized and the colonialists. Discontent on the part of indigenous populations with the cost of living was, therefore, often directed against those immediately available rather than those ultimately responsible.

Another volume in this series has already discussed the nature of colonial urbanization much more completely (see Drakakis-Smith, 1987). But it cannot be emphasized too much that the period from 1850 to 1920 saw a massive restructuring of urban systems. Colonial production may have been based in the countryside but colonial political and economic control was firmly centred on the city. Usually just one or two centres were selected for development, giving rise to the urban primacy which remains characteristic of parts of the region.

The nineteenth century also witnessed the acceleration of segregation within the colonial city as the European presence and dominance became more marked. Small numbers of expatriates thus lived and worked in areas many times larger than those allocated to the indigenous population or to other Asian immigrant groups, establishing an urban pattern that lasted, for the most part, beyond independence and that only disappeared during the rapid urban restructuring of the last decade.

However, such comments give the impression that the colonial patterns and processes established in the nineteenth century continued unabated and unaltered until recently. This is far from being the case, and both subtle and extensive changes occurred throughout the colonial world of Pacific Asia following the First World War (or as some Asians correctly prefer to call it, the Great European Civil War).

Late colonialism

Although no single event can be said to have changed the course of colonialism, let alone a European one such as the First World War, there are clear differences between the periods on either side of that traumatic episode. These contrasts involve new attitudes towards metropolitan relationships with the colonies, changing economic struc-tures and, indeed, changes in the imperial balance of power. Within a single generation the old political world order of 1914 was to fragment almost totally.

Indeed, the fragmentation began, after a fashion, immediately following the Versailles treaty, with Germany being divested of its colonies. However, whilst the major European powers were professing support for the self-determination of the various peoples of the former Austro-Hungarian and Turkish empires, they showed little inclination towards this principal for their own colonial peoples or for those of Germany, who were simply reallocated to other metropolitan powers, primarily Britain.

Nevertheless, there was also a more substantive change within the ethos of colonialism. As Raymond Betts (1985) has put it, the 'heroic' age of imperialism had given way to a more prosaic phase of the development of empire. Here a distinct switch in the attitude and style of colonial government can be detected. The key to this change is the concept of trusteeship which permeated the formation of the League of Nations and in which the well-being and development of colonial peoples was to be elevated to a higher priority. Of course, this did not mean that those subject peoples were actually to participate in this process, far from it: anthropological theory at the time conveniently explained that such progress was impossible within 'backward' and 'traditional' societies.

Colonial government between the wars, therefore, is dominated by bureaucrats striving on behalf of colonial development. Career administrators in both the metropolitan capitals and overseas ran the colonies according to lofty aims but with little real knowledge either of local aspirations or of changing world economics. Indeed, our contemporary image of the colonial period is usually taken from this era, either from the expanding literature or from film: the white-suited gentlemen sipping gin slings, with elegantly dressed wives discussing the problems of keeping good servants, whilst their children are shipped off to boarding school to provide the raw material for Billy Bunter and co. Unsurprisingly, this period was regarded by the privileged few as 'the golden years'. For the colonized populations and, indeed, for the colonies themselves this was far from the truth.

The European wars and depression of the early twentieth century meant that the world economic system had been disrupted. Despite the theoretical views on colonial development, it was still seen as subordinate to the interests of the metropolitan country. Thus, when the basis of Japanese export prosperity, silk production, was drastically undermined by artificial fibres in the late 1920s, the Japanese not only broadened the range of their industrial products but also began to pursue more vigorously the expansion of their Asian colonial empire to provide not only raw materials and fuel but markets too.

In general, however, the European economic depression meant that both government and private capital investment in the colonies was limited. This was exacerbated by the erratic and generally lower prices for colonial export commodities on the world market. Indeed, much of the capital invested in economic activities in the colonies of Pacific Asia during the inter-war period was either American or from the overseas Chinese community that dominated so much local commerce. Thus the largest rubber plantation in the Dutch East Indies was owned by a United States tyre company, whilst 80 per cent of all domestic trade in the colony was controlled by the Chinese, who had used their profits to invest in textile mills.

The declining profitability of primary commodities and increasing urban industrial development is also reflected in the pattern of colonial investment. Figure 2.4 indicates this shift in French Indo-China and reveals in particular the importance placed by the colonial governments on infrastructural development. Schemes to expand road, air and port facilities in particular were the order of the day, no matter how difficult these proved to be. Thus it took almost 40 years to complete the 1,700 kilometres of the Trans Indo-China railroad.

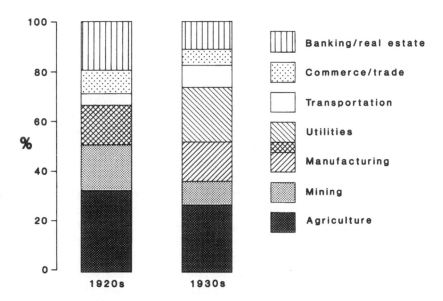

Figure 2.4 French Indo-China: investment during the 1920s and 1930s
Source: Dixon 1991

Despite, or perhaps because of, the precarious economic situation in both Europe and its dependencies, metropolitan and colonial ties grew closer during the inter-war years. Thus by 1930 some 44 per cent of British trade was with the empire. However, the closer ties were not only economic but demographic too, since the European depression had driven many to seek new opportunities in the colonies. Most of the British went to the dominions of Canada, Australia or South Africa but in Pacific Asia the numbers of French and Dutch increased dramatically. In the earlier years many had settled on the land as owners or managers of export-oriented farms or plantations, but by the 1930s most were settling into urban administration or commerce, at levels previously considered not quite appropriate for Europeans.

This deeper penetration of colonial society had two important effects.

First, it closed even more completely the avenues for personal advancement open to the small but growing number of educated persons. Second, it made eventual decolonization a much more difficult process than in neighbouring British colonies. In general, however, the political challenge to the colonial powers was limited during the inter-war period. The oppressed, exploited peasantry and labouring classes, who suffered considerably from the effects of recession, were not organized enough to threaten more than the local, lower echelons of colonialism, whilst the fledgling, elite nationalist movements were too thinly supported. Not until some of the nationalists, such as Ho Chi Minh, returned home from furthering their (political) education in a Europe riven by radical extremism, were the masses and the intelligentsia united under the banner of socialism.

In almost all of the colonies of Pacific Asia, the cities were the focus for much of the change that has been described. Their populations had been swollen by numbers of colonial administrators and migrants, so that by the 1930s in Taiwan, for example, two-thirds of the Japanese population lived in the urban areas, particularly the capital Taipei which alone contained half of this number. But local urban populations were growing too in response to rural hardship, continued land 'reforms' and improved health.

Much of the improvement in urban health resulted from the increased attention given to urban planning, particularly to water and sanitation schemes. In fact, the 1920s and 1930s were marked by the flowering of town planning in Britain and its transfer to the colonies. However, in the colonies two clearly contradictory planning creeds were at work. The first was the growing appreciation of the virtues of the 'garden city' concept, which incorporated pre-existing features of the site into new urban developments; the second principle comprised the imperial urge to demonstrate political power through architectural style and urban structure. In particular, the latter approach tended to ignore what already existed, so that if a ceremonial route was needed, it was bulldozed through whatever indigenous settlement stood in its way.

In general, urban planning of whatever creed was reserved for European areas of the colonial city; the social principles of the garden city movement were thus distorted into a creation of massive green areas (*cordons sanitaires*) between European and non-European populations, largely in the interests of the health of the former. In Pacific Asia most of the indigenous populations were confined to the dilapidated tenements that fringed the European city centre, or to the ramshackle squatter settlements that were beginning to appear on its outskirts. Rarely did the expatriate and Asian populations meet, except in a dominant–subordinate relationship. To most Europeans, their colonial superiority seemed to be unchallenged and secure. The Second World War rapidly brought reality home to them.

Essentially the Second World War within Pacific Asia was fought between its colonial powers: Japan, Britain and the United States. Its principal impact was to destroy European invincibility as Malaya, Singapore, the Dutch East Indies, the Philippines and Burma all fell to Japan within six months. Although most nationalists initially welcomed the Japanese with their slogan of Asia for Asiatics, eventually Japanese assumption of a colonialist role disillusioned all but a few of the indigenous populations. Nevertheless, the humiliation of the French, Dutch and British had opened their eyes to the fundamental weaknesses of the European powers and in all European colonial territories the prospect of the return of the former powers was viewed with distaste and concern.

The actual process of decolonization varied enormously throughout the region and, in fact, cannot be understood without reference to broader changes that were afoot in the world economy. These changes principally revolved around the fact that the United States was clearly the leading world economic power and was keen to penetrate the resources and markets hitherto monopolized by Europeans. The United States, therefore, took the lead in encouraging a climate favouring colonial independence, setting an example by liberating the Philippines in 1945.

Elsewhere the process was much slower, for various reasons. Korea became a battlefield between communism and capitalism; Taiwan was overwhelmed by Kuomintang refugees from the successful communist take-over of the mainland; the Dutch East Indies and Vietnam became embroiled in lengthy and bloody liberation struggles against metropolitan powers attempting to reassert their dominance; other smaller territories such as Brunei, Macao and Hong Kong retained colonial ties for much longer periods.

However, no matter what the status of the various colonies of Pacific Asia, all entered a post-war world which was shrinking rapidly and in which neo-colonial forces tightened the dependent, subordinate relationships with the west. It is worth outlining these forces, alongside the persisting legacies of colonialism, because they provide the backdrop against which the principal elements of the development process can be identified and subsequently discussed within the context of an individual country.

The colonial legacy

Some historians have argued that colonialism was too varied in nature and too short-lived to judge whether its effects have been, on balance, beneficial or detrimental. It is true that even in the relatively limited area of Pacific Asia the complex differences between colonialism over time, across space and between cultures were immense. However, it is impossible and of debatable value to take colonialism out of the global situation of which it was part. Massive changes in the world economic system took place from 1500 onwards and the particular mix of forces that occurred in any one location must be placed in its broader setting. To state that colonialism ended too soon to judge whether mature development of the colonies would have happened is to ignore the fact that an historical perspective indicates that it was already in decline rather than ascendancy by the mid-twentieth century. However, imperialism clearly left a legacy, with which the contemporary states of Asia are still struggling to cope. Let us examine some of the principal components of this legacy.

Politically the national units which emerged in the late 1940s and 1950s were essentially the colonial territorial divisions of the previous century and had limited correlation with pre-colonial structures, with physical geography or with cultural/ethnic patterns. Thus the Malay world of Southeast Asia was divided between Malaysia, Indonesia and the Philippines, whilst Burma comprised the complex mixture of tribes incorporated during the various phases of British annexation. In the climate of mutual distrust which persisted through the 1950s, the interstices of the colonial empires became the object of conflict, such as the Malaysian–Indonesian confrontation over the former buffer states in North Borneo.

Throughout the region, the struggle between capitalism and communism dominated the internal political geography of almost all states, particularly during the 1945–75 period. Not only were states directly created by this conflict, such as Taiwan, North and South Korea, North and South Vietnam, but others suffered long, debilitating internal disputes that soaked up development funds and gave excuses for foreign interference. The United States, in particular, viewed Pacific Asia as the front line in its battle with communism and pumped money, weapons and men into the region for many years, with varying impact. In Vietnam it merely extended and intensified the civil war; in Taiwan and South Korea it laid more solid foundations for economic recovery.

Within these new political units, the demographic legacy of colonialism was also complex. The rate of population growth certainly accelerated. Whether this was the result of a *pax colonia*, improvements in hygiene and health care, or more widespread availability of food, is difficult to assess. Certainly there was a notable shift in population to the major areas of colonial activity, the mines and plantations, and in particular to the colonial cities. The build-up of population pressures (and expectations) following independence added to this urban demographic explosion and has posed some of the most lasting and serious of the social problems facing Pacific Asia over the last 30 years (see Chapter 5).

Another demographic legacy has been the mixing of ethnic and national groups that was the consequence of colonial labour demands: the shift of contracted labour from India to Malaya, for example. In addition, many millions of Chinese migrants have settled all over the region. Most moved in small family units over the years to settle into various forms of commercial activity. British colonialists, in particular, encouraged this process and left most of the local commerce in the hands of such immigrant groups in Malaya, Singapore and Burma. Large numbers of Chinese migrants were, however, not businessmen and were drawn by other labour demands; for example, to the tin mines of Malaya.

The consequence of this racial mixture has varied. In parts of the region assimilation has occurred, but in others, usually where religious cultures are least compatible, the different ethnic groups exist in uneasy, suspicious circumstances ready to erupt when economic deprivation fuels resentment. Chapter 6 will examine this situation in the context of Malaysia, one of the most troubled states of the region in this sense.

In economic terms, the legacy of colonialism has also been substantial. Earlier discussion has highlighted the way in which commercial production became highly specialized into one or two areas creating an economy overreliant on a narrow base for export earnings and vulnerable to fluctuations in the weather, disease or world prices. Moreover, many of these commercial activities were highly concentrated in the more accessible irrigated areas, particularly riverine deltas, giving rise to serious problems of regional and rural imbalance in the pattern of development. The post-colonial response to such spatial inequality has been quite varied in commitment, policies and impact, and forms the basis for the discussion in Chapter 4.

The importance of the colonial period in both the acceleration and redirection of the urbanization process has already been noted. Perhaps what needs stressing too is the fact that most of the urban populations and an enormous amount of the petty capitalist retail and industrial activity were Asian. Many of the problems that have been experienced as a result of the transfer of urban management to indigenous administrators and planners have been dealt with elsewhere in this series (see *The Third World City* by David Drakakis-Smith) and will be summarized here in relation to one of the region's city-states (Chapter 8). This volume will also focus on the ways in which domestic urban capital was used to promote and expand the urban-based industrial change for which the region has become justifiably noted (Chapter 7).

However, analysis of urban-based industrialization and other aspects of contemporary development does not only depend on an understanding of the colonial past, it is also linked to the massive structural changes in the world economy that have occurred since 1970. For some 20 years after the end of the Second World War the economies of most Pacific Asian states remained virtually unchanged. They were still reliant on the export of a narrow range of primary commodities and sought to diversify their economies by a limited range of industries seeking to meet their own domestic market for consumer goods (import substitution). For the most part, these economies were still controlled from outside, through the medium of large multinational corporations or through aid linked to specific development projects, from the former colonial powers to the central government.

In about 1970 all this changed, drastically so in parts of Pacific Asia. Many of the details of what happened will become apparent in later chapters and are fully discussed in Rajesh Chandra's book in this series. In essence what happened was that as the world recession began to bite and profits fell, European and American investors sought out areas where production costs were lower and profitability higher. Pacific Asia has, of course, received a substantial amount of this international investment, although it is very unevenly distributed around the region. Moreover, given domestic investment initiatives in some states, a regional subcentre for manufacturing has been created – with Pacific Asian multinationals increasing not only within the region itself but also in the metropolitan countries too. As always, the benefits of such growth have seldom been distributed on an equal basis and, in particular, we must note the exploitation of Asian labour in this context, particularly female labour (Chapter 9).

It must be appreciated that the series of issues discussed in subsequent chapters is but a selection of the many which characterize the development process in Pacific Asia. Moreover, although generally each is deliberately examined in the setting of a single state, in order to place the background factors (discussed in Chapters 1 and 2) in a real-world, local context, it must be realized that in specific national situations the nature of, and responses to, these issues might well be different. With these *caveats* in mind, let us proceed to examine the way in which some of the region's basic human and physical resources have been managed.

Key ideas

1 Colonialism and imperialism are different concepts. When fused together they produce several major phases of European exploitation of Pacific Asia.
2 The mercantile colonial phase was lengthy but saw relatively little social or political changes.
3 The short industrial colonial phase brought about massive political, economic and social change to the region.
4 The late colonial phase was one of continued exploitation but also of broader development.
5 In the post-colonial period, economic diversification into manufacturing did not emerge until 1970 and was very selective.

3
Physical and human resource management: Malaysia and Papua New Guinea

Introduction

There is a danger in examining the recent changes which have taken place in Pacific Asia of overemphasizing the role of manufacturing. Clearly the growth of export-oriented industrialization has been extremely important, particularly in some states, but in most countries export earnings are still overwhelmingly provided by the export of primary commodities (Figure 3.1). Southeast Asia alone produces 60 per cent of the world's tin, and over 70 per cent of the world's copra, palm oil and rubber.

Many of these export earnings derive from commercial agriculture, a term which covers the production of commodities ranging from basic foods, such as rice, to industrial raw materials such as rubber and copra. These products are grown in a variety of ways ranging from small-scale private production to collectives, large-scale multinationally owned plantations, or state enterprises. Whilst primary commodities of this nature are clearly important in national development, their differing modes of production give rise to a complex set of considerations which are best discussed in the context of overall rural and regional development strategies.

In contrast, the non-agricultural resource exploitation which forms the focus for part of this chapter, employs fewer people but does, of course, have equally important impacts on the economy, environment and on those involved. In essence, an examination of non-agricultural resource exploitation will give an opportunity to scrutinize more specifically the way in which this is managed.

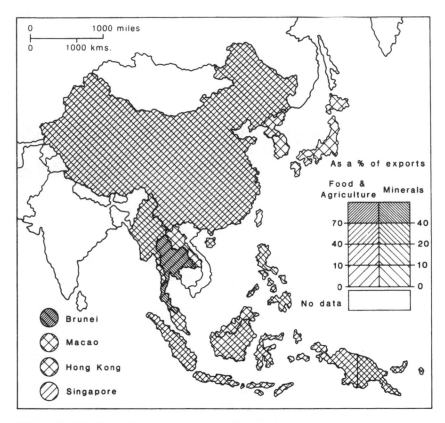

Figure 3.1 Pacific Asia: primary commodity exports

But a nation's resources are not only physical (mineral, agricultural products, energy, etc.), they are also human and in Pacific Asia, in particular, it is people who have proved to be the fundamental basis for prosperity. A glance at Figure 3.1 indicates clearly that the four leading industrial nations of the region have few natural resources and yet their per capita incomes are amongst the highest in the Third World.

Human resources are, of course, as exploited as any other type of resources; this will form an important focus of Chapter 7. In the same context, some would claim that human resources need to be managed properly if maximum efficiency is to be achieved. One would hope that this means that individuals are given the opportunity to realize their own potential. Unfortunately, human resource management is not normally considered in such generous terms and when human welfare, in terms of education or health, is drawn into development planning it is usually in the context of improving the quality of the labour force rather than the quality of life.

This dimension of 'resource management' will form the second focus for this chapter. Although the subject matter of each section appears to be quite different, very similar moral and ethical considerations are at issue. Just who is development for? Who gains and what price is paid?

Non-agricultural resource exploitation: some background considerations

Natural resource exploitation can take varied forms but most can be clustered into two main types. First, is enclave exploitation in which spatially limited activities occur, often in isolated locations. A good example of this type of project has occurred in Papua New Guinea where the huge Ok Tedi copper mine is situated in a region so isolated (near to the border with West Irian) that it took five years simply to provide suitable access to the site before any mining could begin (see Case study C). The other main type of exploitation comprises large-scale, extensive activities such as tin mining or timber extraction. It is often assumed that the environmental and social impact of these projects has a more extensive impact than enclave activity but this is not the case since the latter, although more localized, often occurs in more remote, traditional areas where the sudden infusion of new technology and new values can have catastrophic consequences.

The sheer scale of most kinds of resource exploitation means that they require considerable capital input, particularly if expensive technology is necessary. Many Third World governments cannot afford such costs on their own and are reliant upon large multinational corporations (MNCs) to fund, organize and undertake these operations. In most cases MNCs operate in their own interests, but under certain circumstances there is a close link between state and company policy and operators. This occurs extensively in Pacific Asia through the operations of Japanese MNCs.

Case study C

Papua New Guinea: Ok Tedi bares its soul

Mount Fubilan in the remote, jungled interior of Papua New Guinea is a mining corporation's dream. It comprises a huge copper core with gold-bearing deposits on top. The smaller-scale exploitation of the gold since 1984 has financed the more extensive preparations needed to extract the copper.

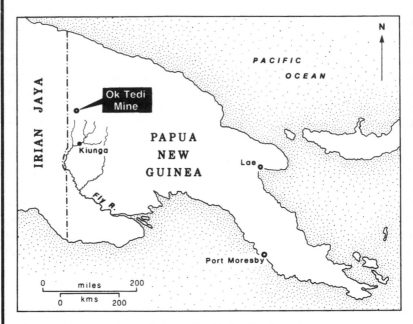

Figure C.1 Papua New Guinea: Ok Tedi Mine location

The region was unexplored until the early 1960s and the ore deposits were discovered soon after in 1968; it has taken 20 years to begin production because of the inaccessibility of the site (Figure C.1). Although the Fly River is navigable to Kiunga, there was no route over the next 100 miles of dense, perpetually wet forest. With great difficulty (and loss of life because of constant

Case study C *(continued)*

low cloud), heavy machinery was airlifted in and literally cut its own way out to the river. Total investment costs amounted to US$ 1.4 billion but the richness of the ore and cheapness of local labour make production costs only about half of those elsewhere in the world.

The MNC involved in Ok Tedi Mining is a consortium of Australian (30 per cent), United States (30 per cent), West German (20 per cent) and Papua New Guinea government (20 per cent) interests. Long-term contracts already signed commit about 80 per cent of the expected output over the next ten years or so to West Germany, Japan and South Korea. Copper production proper began in 1988 and with world prices for copper holding up well, it is expected that 150,000 tonnes can be produced each year.

Within 16 or 17 years about 22 million tonnes of ore and waste will be torn out of the area, turning what was a mountain into a 400-metre deep hole in the ground. What will the environmental impact be? How will the local population adjust to, and cope with, the various demands for labour, land and services? Whilst the economic prospects have been widely discussed, little information on the ecological impact is available, even though this has all happened before in Papua New Guinea at Bougainville.

Source: Drawn from Michael Malik 'Moving the mountain', *Far Eastern Economic Review*, 28 July 1988.

One general problem which has affected non-agricultural resource exploitation in recent years has been falling world prices. Although the 1970s were marked by OPEC successfully establishing an oil producers' cartel which monopolized the production and therefore the price of oil, similar attempts by other Third World commodity producers were unsuccessful. There are a variety of reasons for this, including the desperate need of some countries for immediate rather than long-term revenues and their consequent undermining of producer solidarity by arranging separate deals to increase their sales. In addition, many of the commodity producers were heavily reliant upon imports from developed countries for further economic diversification and growth (tractors, fertilizers, machinery and the like); they were thus vulnerable to retaliatory measures.

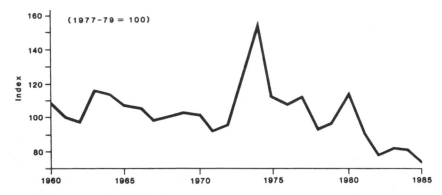

Figure 3.2 Commodity price trends
Source: World Resources (1987)

The result of this failure to capitalize on monopoly of supply has been a steady downturn in world commodity prices (Figure 3.2). But dwindling prices have not necessarily resulted in a fall in production. The great majority of developing nations have no alternative sources of revenue to which they can turn; they are totally reliant on a few export commodities for their export earnings. The consequence in some countries has been an acceleration of production. If it takes twice as much tin or timber to obtain the necessary revenue to pay for the import of one new tractor or one sophisticated tank, then the developing countries seem to have no choice. One alternative is to raise productivity per capita. This is usually done by increasing output whilst holding production costs down; invariably this means wages, and presupposes collaboration between MNCs and most governments. This is exactly what has happened and the following sections examine the repercussions of such collaboration and acceleration of production for the environment and the labour force.

Malaysia: some environmental impacts of non-agricultural exploitation resource

Introduction

Malaysia is one of the more prosperous nations in Pacific Asia but, despite the recent rapid growth of manufacturing output, its prosperity still derives extensively from the export of primary commodities. Minerals and oil account for 25 per cent of exports, with agricultural products another 36 per cent. As Chapter 6 will reveal in far greater depth, the first extensive resource extraction was that of tin in the late nineteenth century. The exploitation of this non-renewable resource was accompanied by extensive rail and road construction which, in turn, paved the way for subsequent land clearances and the development of large-scale plantation agriculture of commercial products, such as rubber and palm oil.

The value of these commodities to the Malaysian economy has ensured that the government has supervised the renewable resource exploitation of commercial agriculture very closely, both at plantation and small-scale levels. Research institutes have been established, regular replanting is encouraged, and advice and subsidies are readily available. Elsewhere, however, the picture is not so positive and to a certain extent this attention to the needs of commercial agriculture has clashed with, and is in sharp contrast to, the management of another of Malaysia's principal renewable, non-agriculture resources, i.e. timber.

It is well known that the world's tropical rainforests are under increasing threat (see Avijit Gupta's book in this series, *Ecology and Development in the Third World*), and Southeast Asia is no exception. Malaysia's forest resources, like all others, have been reduced by land clearance, increasingly desperate fuelwood cutting, depredations from mining and inadequately controlled timber extraction. Whilst the last of these will form the focus for the following section, it is worth noting that sheer population pressure is often the primary cause behind many of the problems listed. For example, in the Philippines the reduction of 30 million hectares of hardwood forests to just 1 million hectares is said to be the result of intensified shifting cultivation in the country's less developed areas. The 'slash and burn' approach has not just depleted the forest but virtually annihilated it.

Commercial logging in Malaysia

Malaysia, of course, has inherited many serious environmental problems from resource exploitation, the most extensive of which is probably the barren, sandy tin tailings (waste) which cover more than 80,000 hectares of what is a small country. Together with poor management of industrial processing, such as rubber production, this has resulted in the drastic pollution of some 60 rivers, 40 of which have been declared officially dead of aquatic life.

Unfortunately Malaysia seems not to have learned its lesson and timber extraction has been equally profligate in recent years. Increasing demand for tropical hardwoods has seen vast concessionary areas given over to multinational corporations, usually Japanese. Malaysia, together with the Philippines and Indonesia, has provided the brunt of Asia's timber production over the last 20 years (Figure 3.3), and by the mid-1980s its roundwood and sawnwood exports amounted to well over two-thirds of the total for the whole of the continent. Economically this has proved extremely valuable at a time when other commodity prices are falling, and timber exports now bring in more than tin, palm oil or rubber. Ecologically the pace of extraction has all the makings of a disaster since it is estimated that at present rates Malaysia's tropical rainforests will be obliterated by the end of this century.

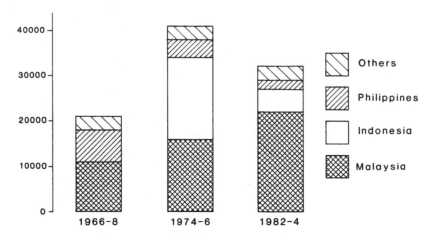

Figure 3.3 Pacific Asia: timber exports

It may be true, as the government claims, that logging is selective in Malaysia, meaning that only certain trees are to be felled. But gaining access to particular species often means that a 10 per cent selection results in the loss of half the trees, whilst another 20 per cent are damaged (Plate 3.1). But the impact on the environment is greater than just the loss of trees, valuable though they are, since the gradual shifting of logging activities downslope and upslope (the cropping limit was raised from 300 metres to 900 metres by the government) has drastically affected the neighbouring environments. Downslope it is estimated that

60 per cent of coastal mangroves have been lost during the 1980s. This means a loss of fuelwood, and of feeding grounds for fish and shellfish which in turn means less food and a loss of livelihood for coastal villagers. Upslope, of course, logging means soil erosion and consequent silting of rivers, as well as a threat to the habitat of many forest tribal groups (see Case study D).

Plate 3.1 Sabah: logging access and its consequences. Even in the initial access stages huge swathes of the forest are lost.

At the heart of these problems in Malaysia lie split (and abandoned) responsibilities. Land and its management are in theory the responsibility of the individual states that make up the Malaysia Federation but the profits from other commercial resource exploitation, such as tin mining or rubber plantation, belong to the companies involved and the federal government (via taxes, licences, etc.). When it comes to logging, the less developed states are, therefore, unable or unwilling to intervene because timber is their principal source of income. To give an example, in Pahang in 1978 the last extensive block of lowland rainforest was scheduled for gazetting (official declaration) by the federal government as a forest reserve but the state government granted logging rights, claiming that human welfare (through employment) was more important than conservation. Subsequent investigation revealed that the humans whose welfare benefited most were a small group of businessmen and corrupt officials. There was a great public outcry but this served only to accelerate the activities of the logging company before the gazetting deadline.

Since independence, development in Malaysia has been characterized by a policy of 'develop now and tidy up later'. This may be possible with regard to environmental pollution, although it does not condone the lack of responsibility, such a policy is not possible with a disappearing result. It was only in the late 1980s, as a result of increased pressure by world environmental organizations, that the Malaysian government belatedly took action. Hitherto the principal pressures had been exerted by the multinational corporations involved. The outcome, as far as logging is concerned, has been a ban on timber exports imposed in 1989. Clearly this is not a philanthropic ban: the argument of the conservationists which most appealed to the Malaysian government is that timber is a renewable resource, but that only through more careful regulation will it continue to furnish export returns in the future.

Case study D

At loggerheads in Sarawak

Sarawak is one of the constituent states of the Federation of Malaysia, situated on the northwest coast of the island of Borneo. In addition to the rich tropical forests, it contains a variety of indigenous peoples collectively known as Dayaks. Most of the Dayaks live in longhouses and practise a shifting agriculture in which the forest is cleared only for one or two years before being allowed to regenerate. Some tribes, such as the Penan, still survive by hunting and gathering but have found this increasingly difficult as logging companies have invaded their lands.

Over the last decade Malaysia has been the subject of intensive, one might say frantic, logging activities – most of which have occurred in the East Malysian states of Sarawak and Sabah. By 1985 one-third of Sarawak's forest area had been logged; another

---

Case study D *(continued)*

third will be gone by the end of the century, the concessions have already been given to the timber companies. Most of the timber is exported as sawn logs, so the state does not even benefit from associated industrial development. Japan takes over half of the exports.

Logging and politics are inextricably mixed in East Malaysia. The Minister of Environment and Tourism, for example, is also the head of one of Sarawak's largest timber companies; whilst the leaders of the two main political parties are alleged to belong to commercial groups which between them hold concessions covering 30 per cent of the entire forested area of Sarawak. The uncontrolled logging has not only permanently destroyed much of Sarawak's forest resources, it has led to soil erosion, ruined more than half of the state's rivers, and wiped out many of the forest fauna. In short, it has deprived the Penan and Dayaks not only of their livelihood but of their lifestyle.

In 1987, supported by state, national and international organizations, the Penan issued an appeal:

> Stop destroying the forest or we will be forced to protect it. . . . You took advantage of our trusting nature and cheated us into unfair deals. . . . We want back our ancestral land, the land we live off. We can use it in a wiser way.

The appeal was ignored, so the Penan began to set up barriers across access roads, bringing work to a halt in dozens of logging camps. A group of 12 leaders travelled to Kuala Lumpur to complain of the unsympathetic response from the politicians in their own state. A few months later graphic evidence of this appeared when the blockades were dismantled and 42 tribespeople were arrested and charged with, amongst other things, 'unlawful occupation of state lands'. The Penan, of course, regard the lands as traditionally their own.

The case was taken up by Survival International, a British-based organization which protects the rights of threatened tribal peoples. Together with other similar groups, it began to call for an international boycott of Malaysian timber. The state government

Case study D *(continued)*

dug in its heels and vowed no concessions. In May 1988 the Dayaks re-erected their barricades. This was soon followed by a resolution from the European Parliament calling for a halt to imports of hardwoods from Sarawak. Eventually the federal government stepped in and declared a ban on exports, much to the annoyance of state officials.

Source: This case study was compiled from various reports from the *Far Eastern Economic Review* and Survival International. A slide and tape presentation of the Sarawak situation may be obtained from SI at 310 Edgware Road, London W2 1DY.

Papua New Guinea: the fusion of physical and human resource exploitation and reaction

Non-agricultural resource exploitation in Pacific Asia, as elsewhere in the Third World, not only has environmental repercussions. There are also too many occasions when there are adverse effects on the indigenous human population. Much of this is related to the labour needs of resource extraction. Most mining or forestry activities, even those in remote areas, require capital-intensive investment and yet, paradoxically, also demand large numbers of unskilled and semi-skilled labour. It might reasonably be imagined that as such activities occur in remote and difficult areas, wage rates would need to be high in order to attract the necessary workforce. However, this is not the case and irrespective of local population densities (and therefore the size of the potential labour pool), wages for indigenous employees are normally uniformly low.

The reasons for this were twofold. First, the quality of much local labour was low – so unskilled and scarce in some areas that workers had to be brought in from other regions or even from overseas. Second, initial labour recruitment in 'traditional' areas tended to be on a part-time basis, for very short periods, of men whose primary loyalties, responsibilities and economic support remained in the traditional sector of the rural economy.

Such men were content to join the labour force on a casual basis and for low pay because they sought remuneration for specific targets, i.e. to pay the annual poll tax or to meet family wedding expenses. On their part, the companies paid low wages in the full knowledge that they were not supporting the men's families. Thus the welfare component of the wage, such as insurance, superannuation or housing allowances, was non-existent. The result was a very low wage bill, admittedly for unskilled, unmotivated labour with a high turnover.

The problems start once the workers begin to change their lifestyle and, consequently, their material and financial needs; in other words, once they became more committed to a non-agricultural employment and remuneration. This change in personal lifestyles means little to the mining company which continues to pay low wages, often claiming that falling world prices (and profits) preclude any increase in wage bills. In most cases, however, mining companies have a considerable amount of leeway or slack in terms of their profit, some of which could be transformed into higher wages before operations become genuinely unprofitable. However, the threat of cessation and withdrawal is often used to coerce national and provincial governments into controlling labour unrest and agitation for fairer wages/conditions. Such was the case in Papua New Guinea at the Bougainville copper mine.

Bougainville: the evolution of labour conflict

The island and province of Bougainville (or North Solomons) was, until the twentieth century, a place of subsistence affluence where small tribal groups lived in harmony with their environment but in a state of perpetual antagonism with their neighbours. Contrary to what one might expect from its name, the island was initially colonized in the late nineteenth century by Germans who had gradually built up some copra plantations by 1914.

After the First World War the island became mandated to Australia who administered Papua New Guinea as a colony until 1975, by which time Bougainville had a population of just under 100,000. Development under Australian administration continued to be centred on copra with labour being forced into the plantations by the imposition of a head tax. Meanwhile, the activities of rural Christian missionary factions served only to maintain and intensify intergroup violence.

Indigenous resentment towards Europeans was simmering slowly when in 1964 Bougainville was found to be rich in copper, gold and silver. The subsequent exploitation of those deposits caused massive acceleration in the development and exploitative processes. This is clearly reflected in the rise in the urban population from just 740 to over 14,000 between 1966 and 1971.

Some of this new urban population was European but many were workers recruited from other parts of Papua New Guinea. A complicated resentment began to build up in the local population as the mining company was permitted to expropriate land for its activities by the national government. The local population thus felt exploited both by foreigners and by their central government, and separatist agitation emerged. This was further complicated by hostile feelings towards the Niuguinian mineworkers recruited from elsewhere in the country.

These various factors served to unite Bougainvilleans as never before at a time when the country as a whole was moving towards indepen-dence and the prospective national government wished to retain political unity. It resulted in 1973 in the province receiving substantial royalties from the mining operations which were used to fund local rural development projects. In addition, people living in the vicinity of the copper mine received compensation payments for expropriated land, loss of crops, pollution and the like. Yet this was only part of the struggle, exploitation of a different kind was occurring within the mining company itself.

Alex Mamak and Dick Bedford (1979) have written a graphic account and analysis of one particular incident which encapsulates many of the features of exploitation and workers' reactions to this. The incident occurred in 1975 some three years after copper production commenced in Bougainville and had at its heart a call for higher wages and better working conditions. During the first meeting between the company management and union officials, the police had been called in to disperse about 1,000 mineworkers who had gathered to show solidarity with their representatives. Clashes occurred and for two days there was a running battle between workers and police which ended in mass arrests and massive damage to company property.

Although press coverage played it down, racial antagonism was an important dimension of the conflict. The reason for this was straight-forward: the best incomes, housing and standards of living were enjoyed by the company's white expatriate workers. Labour exploitation was seen by the workers as having a clear ethnic dimension.

Mamak and Bedford produced information on average earnings which clearly supported this interpretation of the situation (Figure 3.4). Although there was a class dimension to the resentment felt by the bulk of the local workers, in general the fact that there were separate local and expatriate pay scales for both white-collar and blue-collar workers, even for the same jobs, meant that ethnic solidarity largely bridged class differences.

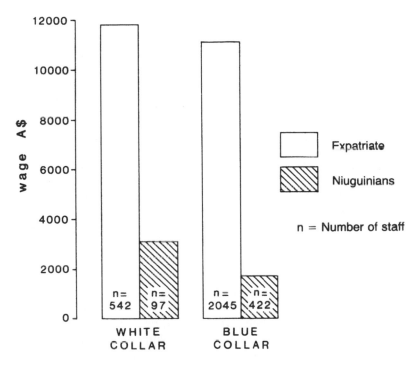

Figure 3.4 Papua New Guinea: wage differentials in Bougainville copper mining
Source: Mamak and Belford, 1979

Much of the public discussion and interpretation of the events of May 1975 stressed class cleavage and resentment between local employees, but for the union and its members the dispute was clearly caused by wage exploitation and repression of its local labour force. In the end the union's solidarity won the day, not only in terms of improved wages but also in terms of gaining greater respect for, and consideration of, the indigenous workforce in a broader sense. Wage differentials have closed somewhat, more Niuguinians are in senior positions and, more importantly, their advice is sought on industrial relations.

But such progress was too little, too late to halt the deterioration in relations between islanders and expatriates and these declined further in the 1980s to the point in 1989 when another Bougainville secession movement resulted in the government declaring a state of emergency and sending in 2,000 troops. The rebels not only want to run their own province but also want to force out the main multinational mining company.

This episode in one of the more remote parts of Pacific Asia may seem to have heavily local characteristics but the basic issues of labour exploitation and ethnic discrimination are common to much non-agricultural resource exploitation throughout the Third World, no matter whether the resource in question is copper, oil or timber.

Human resource potential: exploitation or realization?

Thus far our discussion has concentrated on resource mismanagement rather than management and on the way this has led to exploitation of both the environment and humans. In many parts of Pacific Asia it is human rather than physical resources on which sustained economic growth depends. The detailed ways in which this has affected rural and urban development will be discussed in later chapters but it is important to appreciate that an improvement in the quality of life at all levels, from the households to the nation, depends considerably on helping individuals to realize their own potential.

Figure 3.5 illustrates the main components that affect the nature of human resource potential. There are many others, such as climate and culture, but these are less amenable to adjustment. As the diagram clearly indicates, there is a web of interrelationships between these factors. For example, low incomes mean inadequate diets and health care which, in turn, mean that adults leave the workforce early, children are therefore kept out of school and grow up uneducated and unable to appreciate new ideas on health care or nutrition, and have as limited a range of employment opportunities as their parents. In short, the whole cycle of deprivation is perpetuated.

Any effective change to this situation will be long term. Although some programmes, such as mass vaccination campaigns, can produce spectacular short-term gains, these tend to be the exception. Improvement within the household is a gradual affair; children are more likely to be educated at secondary level if their parents have already been to primary school. Full impact is measured in years rather than months.

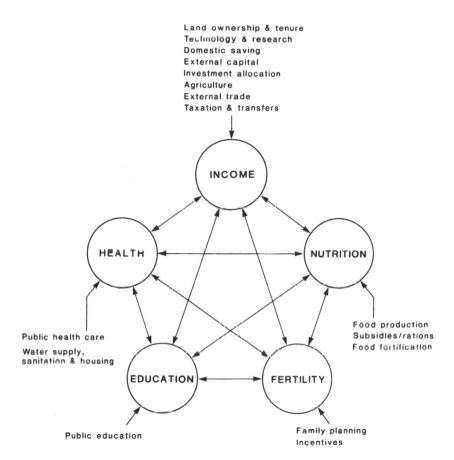

Figure 3.5 The human reosurce network
Source: World Bank Development Report 1979

A second point to remember about improvement in human resource potential is that it is heavily dependent on government investment and support. Poor families can do little to improve their condition because they are poor. Unfortunately the long-term, less tangible nature of welfare improvements make them less popular with most governments than investment in economic enterprises. Measures which are often regarded as social overheads are usually the first to be sacrificed during recessionary periods when government resources are stretched. The only programmes that tend to persist are those related to fertility control, which are pursued instead of other types of social investment rather than in parallel with them.

One final but crucial point which must be borne in mind in assessing programmes related to human resource development is their *raison d'être*. The rationale behind improvements in education, health care or nutrition ought to be the creation of an environment in which individuals can realize their own potential for the benefit of themselves, their family and the broader community in which they live. All too often, however, this list is reversed and the objective behind social investment is simply to improve the quality of the labour force.

It is with this cautionary note over the objective of social welfare investment (which is usually so rare that enthusiasm overwhelms judgement), that we will proceed to examine three of the five elements identified in Figure 3.5. The question of poverty and improving incomes is addressed directly and indirectly throughout this volume; demo-graphic considerations are discussed in Chapter 5; so that leaves nutrition, education and health care to be briefly reviewed in the context of Pacific Asia. Many of the relevant statistical data for the region can be found in Table 1.1.

Education

During the 1950s and 1960s educational investment in Pacific Asia was strongly influenced by overall development strategies which recom-mended modernization along western lines. As a result, primary educa-tion was neglected in favour of specialist training at secondary or tertiary levels. Whilst this has changed in recent years, post-secondary investment still heavily outweighs primary spending by eight to one. The effect in some countries has been to produce an excess of graduates who are absorbed into an inflated government bureaucracy in order to forestall idle minds turning into political malcontents.

Nevertheless, as Table 1.1 indicates, primary education is now the rule rather than the exception over most of Pacific Asia, and secondary enrolments are increasing rapidly. Adult literacy rates are thus impress-ive compared to other parts of the Third World, although women are still disadvantaged in this respect. Increased access to education has a positive effect on equity, since it improves the income-earning capacity of the poor directly and also lays the foundation for indirect improve-ments, for example, by leading to increased responsiveness to new ideas on nutrition or health care.

However, investment in primary education is often quite different from ensuring the poor benefit accordingly. One major problem in this respect is accessibility. Access is not only a physical problem for children in rural areas, especially if these are thinly populated and remote, it is also an economic problem when families cannot afford the indirect costs of education, such as school books, uniforms, pens and pencils. Some 95 per cent of recurrent expenditure on primary education in Pacific Asia goes on teachers' salaries, so that other subsidies are not available. It is an indication of how highly Asian families rate education that many household sacrifices are made to this end. For example, rural families may sell land; or elder children, usually daughters, may take on factory employment in order to keep younger siblings at school (see Case study R in Chapter 9).

As this suggests, reduced accessibility in education is likely to be gender biased against women. For many families in Pacific Asia, investment of limited resources in the education of girls seems not to be worthwhile as they join their husband's family after marriage. And yet it is education of women more so than men which pays handsome dividends for the well-being of the family since it becomes better fed and healthier, to its ultimate economic advantages. But the principal benefit is to expand the role of women in the decision-making process within the household, giving them more self-respect and enabling them to realize more fully their own potential as human beings.

The links between education and other human resource components is clear, but they can work negatively as well as positively. If children are unhealthy or underfed, they under-achieve at school. Investment in education alone is, therefore, to reduce its effectiveness. Parallel programmes in other aspects of social welfare are required.

Food and nutrition

Assessments of world nutritional problems have changed drastically over the last two decades. In the past the primary difficulty was considered to be dietary shortages of important minerals and vitamins. Now the problem is seen simply as a reflection of not enough food, i.e. with undernutrition rather than malnutrition. Another notable change has been the switch away from pessimism about population – food trends, which in the 1960s gave rise to a neo-Malthusian clamour for fertility control – to a more 'optimistic' interpretation of events. An unexpectedly early reduction in population growth rates and a rise in world food production have reduced current estimates of the Third World population affected by undernutrition to about one-quarter of the total.

Within this global situation, Pacific Asia fares reasonably well, particularly as cereal production has risen relatively rapidly (see Chapter 4). However, below this aggregate level important variations are revealed. Thailand and the Philippines, for example, have experienced rapid population growth which has put pressure on food intake despite expanded production (Table 1.1). But it is below the national level that the real problems are revealed, since nutritional well-being is strongly correlated with income.

The situation is particularly difficult for the urban poor whose opportunities for subsistence cultivation are few, particularly in the largest cities. Many of the urban poor spend more than three-quarters of their meagre income on food and must sacrifice other basic needs in order to survive. As a result most live in squatter settlements where housing costs are low, or keep children out of school so that they can earn some money to add to the household budget. The situation is often exacerbated by cultural prejudice and/or ignorance within the household. This particularly shows itself in the bias in favour of males. When there is insufficient food, it is usually women and girls who go without. It is no coincidence that when we see television coverage of eminent persons visiting Asian hospitals, the undernourished children are almost all girls.

Internal responses to these nutritional problems have been disappointing. Increased agricultural production has occurred almost everywhere in Pacific Asia but this has not necessarily benefited the poor. Much of the increased production is of non-food crops and many countries gain considerable export earnings from agriculture whilst their peasants go undernourished. It is in this context that the results of the 'green revolution' have been most disappointing.

The paradoxical position for most of Pacific Asia is that, despite increased agricultural output, food imports have increased almost everywhere (Figure 3.6). This is not just a matter of accelerated population growth. It is linked to the priority given to export cropping and to the replacement of traditional values by western ones so that dietary preferences change and indigenous foods become less desirable. This, of course, is a process which has been encouraged and accelerated by food multinationals. One of the most notorious instances in this context was the diffusion of powdered milk throughout Pacific Asia by a European firm which had purchased and processed the EEC milk lake. The substitution of this product for mother's milk (undertaken through misleading advice from quasi-medical teams) resulted in an escalation of both company profits and infant mortality rates.

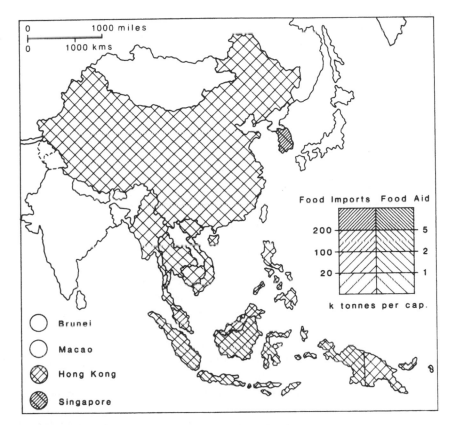

Figure 3.6 Pacific Asia: food imports and food aid

It is nutrition, perhaps more than any other welfare issue, which illustrates so well the interdependent nature of human resource improvement. Essentially, the nutrition of the poor will not improve until education, particularly for women, becomes more readily available; until health care facilities are upgraded; and until there are fewer mouths to feed. Moreover, these need to be integrated. Recent research has shown that raising income levels alone does not always bring about improvement in household nutrition levels, because of other factors, principally related to ignorance about proper diets or the benefits of traditional crops compared to westernized foods. Nevertheless, it is also true that governments themselves often seem to need educating as much as the poor in the ways to improve nutritional intake.

Health care

The contributory factors to health deficiencies in Pacific Asia are many and varied. They range from the regional environmental factors, such as the tropical climate over much of the region which enables disease vectors to remain active all year, to more local environmental factors of pollution (see Case study E) and poor sanitation and water supply. In addition, the inadequate provision of related basic needs, such as low levels of education and undernutrition, increase vulnerability to diseases of all kinds.

Case study E

Hong Kong: not so fragrant harbour

Hong Kong has one of the most enviable economic growth rates in the world and its enormous production comes from a tiny area of just over 1,000 square kilometres. Its thousands of factories and millions of people have sprawled from their original site around Victoria Harbour to encompass almost all of the main island and most of the New Territories of the mainland. Whilst this growth was occurring, the government had many priorities for its public expenditure – housing, roads, utilities. Sewerage was near the bottom of the list and the human and industrial waste was simply poured into the sea; in short, the solution to pollution was seen as dilution. The trouble was that little of the waste was treated and the outflow pipes were not long enough, merely dumping the waste a short distance offshore.

Rather unsurprisingly this has resulted in massive water pollution. Over 1.6 million cubic metres of waste flow into Victoria Harbour each day and microbiologist John Hodgkiss has found that the water at the popular Repulse Bay resort was 52 times more polluted than European safety standards. But the worst affected body of water is Tolo Harbour, in the northeast of the colony. In the 1970s this was a relatively underdeveloped area with vegetable farming being the most intensive form of development (Plate E.1). Now the area contains almost a million people in various new town developments, plus their associated industrial zones.

During this development the limited legislation related to waste

Case study E *(continued)*

Plates E.1 and E.2 Hong Kong: Shatin Valley, past and present. Views of a valley in the New Territory before and after it became the site of a new town.

Case study E *(continued)*

disposal was the responsibility of the Buildings and Lands Department, but despite massive flouting of lease regulations no factory has yet been taken to court. Belatedly it has now set up a pollution unit which has begun to identify factories illegally dumping waste, but for Tolo Harbour this was not regarded as sufficient and the government set up an Environmental Protection Department in 1986, with a budget of HK$90 million and a staff of 350.

Although several substantive new anti-pollution measures have been passed, the impact has still been quite limited. First, the anti-pollution laws that have been passed are spatially restricted, those relating to water, for example, were initially applicable only to Tolo Harbour. Second, the legislation applies only to new industries; existing factories are exempt and are to be subject only to the pre-existing complex of ineffective regulations. Third, there is insufficient information on which to base future legislation since industry for the most part goes unmonitored. Ironically, Hong Kong's water treatment facilities are inadequate and would be unable to cope with a fully compliant industrial sector.

In the face of increasing criticism of its spineless subordination of environmental interests to those of industrial profit margins, the Hong Kong government responded with new legislation. Unfortunately this did not challenge the major offenders, but rather the colony's 6,000 small-animal farmers who were confronted with tough new laws on animal-waste disposal. This was useful legislation but tackled only a small fraction of the problem and was aimed at a group of people who carry little political clout. However, they did carry plenty of physical clout as the governor, Sir David Wilson, found as he tried to leave the Legislative Assembly in his official limousine. No doubt he was grateful for small mercies – a mob of industrialists might have done much more than dent his car or his self-esteem, they could have threatened the surplus remitted to London. This, after all, is Hong Kong's *raison d'être* and apparently no environmental protection measures must threaten it.

Within Pacific Asia as a whole, children tend to be more vulnerable than adults and, despite the relative prosperity of the region in comparison

to the rest of the Third World, infant mortality rates are still high compared to those of developed countries, with the exception of Singapore
and Hong Kong (Table 1.1). The main contributory factors are the pre-
valence of intestinal diseases and respiratory infections, the impact of
which is worsened by undernutrition and squalid living environments.

Within individual countries, rural areas tend to display worse health
data than urban areas. This is because most investment in health is spent
on expensive hospitals and clinics which tend to be located in large
cities. However, access to these facilities is not uniformly equal because
of cost; moreover given that the urban poor tend to live in more squalid
conditions and have fewer opportunities to grow subsistence food, their
health tends on the whole to be worse than that of the rural poor.

Within individual settlements, therefore, there are often sharp
contrasts between low-income districts and other parts of the city. Data
to support these assertions are frequently difficult to obtain but one
survey in Manila, for example, revealed that in squatter settlements the
incidence of anaemia and diarrhoea was double that for the city as a
whole, whilst for tuberculosis the differential was nine times.

In the few studies that have been undertaken of different poor
communities within cities, contrasting health problems have also
emerged relating to local environments. For example, in Singapore 75
per cent of squatters had intestinal parasites compared to 32 per cent
of those living in low-cost government flats, largely because of the
difference in housing quality. However, those living in flats were more
likely to suffer from illnesses related to undernutrition as families
sacrificed food purchases to meet rental demands.

This lack of appreciation of the particular problems facing the urban
poor has meant that the most fundamental change in health care policy
of the 1980s, i.e. the shift to primary health care, has been oriented
almost completely at rural areas. Primary health care has risen to
prominence since the realization by world health authorities that the
transfer of medical mass technology to the Third World has reached a
plateau in terms of its effect. Indeed, as with malaria, pesticide-resistant
vector strains have emerged over the years, thus reducing the impact.
The next phase is to seek to improve the health circumstances and
health behaviour of the family itself. The most crucial social investment
in this respect is the provision of clean, safe water (see Case study F).
But reticulation of water (and sewerage) systems is very expensive
compared to immunization programmes and in the meantime the World
Health Organization and UNICEF are emphasizing primary health care

programmes through the medium of community paramedics, trained to advise in health care, fertility, nutrition and related matters but referring more serious cases to back-up services.

There is considerable scope within this arrangement for the incorporation of traditional health care systems. These used to be derided as backward or based on sorcery, but appreciation of the effectiveness and evolutionary nature of traditional Chinese health practices has finally opened many western minds.

Case study F

Water and health: some major waterborne diseases common in Pacific Asia

Cholera: highly infectious and sometimes fatal, marked by diarrhoea and other gastrointestinal symptoms. Incubation period of only three days.

Typhoid: also very infectious and sometimes fatal, marked by diarrhoea, fever, headaches and intestinal inflammation. Incubation may take several weeks and causes problems in identifying source of infection. Infected persons may continue to be carriers even after recovery.

Dysentery: there are several different types of dysentery, one of which is a major cause of death amongst infants where sanitation is inadequate. Spread through faeces and with a short incubation period of only four days.

Polio: a crippling viral disease attacking the central nervous system, causing paralysis of the lower limbs, particularly in children.

Hepatitis: highly infectious and frequently fatal viral disease affecting the liver, usually transmitted through consumption of contaminated water or food. Symptoms may include coma, nausea, and physical and mental debilitation for lengthy periods following the initial attack.

Low-cost purification

It is estimated that water-related diseases are responsible for 80 per cent of public health problems. Recent research has concentrated

---

Case Study F *(continued)*

on simple, low-cost methods of water purification to enable poor communities to remove most of the worst bacteria and viruses from the water. One important project for Pacific Asia has shown that the ash from burning rice husks (a common waste product in the region) can be mixed with cement and water to produce inexpensive water filters. It is this type of research which holds most promise for the poor who are worst affected by water-related health problems.

Source: Abstracted from 'Fresh Water: the Human Imperative', *Searching*, International Development Research Centre, Ottawa, Canada, 1989.

Whilst the emphasis on primary health care is to be welcomed since it directs attention to the specific needs of the poor, and also draws into the system many other vital components of that seamless web of interacting elements (education, fertility, nutrition), we must be aware that it can be used as a cheap substitute for more fundamental (and costly) social investments into schools, clinics, water supply systems and the like. Indeed, there is some evidence in Pacific Asia that urban primary health care schemes are simply showpiece ventures injected into limited areas of capital cities and designed to impress international visitors.

This rather cynical interpretation is, unfortunately, only too accurate when it comes to investment in welfare programmes designed to allow the poor to realize their own resource potential. The principal priorities for national development strategies tend to be political and economic rather than social, as subsequent chapters will reveal.

Key ideas

1 Resource exploitation often has considerable environmental impacts.
2 Human resources are often exploited rather than developed. Concern is with the quality of labour rather than the quality of life.
3 Basic human needs are closely interdependent. Improvements in one will not be as effective as integrated improvements in all.

4
Rural and regional development

Introduction

Although Pacific Asia understandably has a reputation for its urban industrial development, it is nevertheless the case that most people in the region still live and work in rural areas (Figure 4.1). More people are classified as rural than work in agriculture (or fisheries), however. This is mainly due to employment in rural non-farm occupations; but in the more urbanized countries it is also a symptom of rural commuting to urban areas and the advance of suburbanization. There is, of course, substantial variation within Pacific Asia in the importance of the rural sector and this cuts across political affiliations. Thus there are both capitalist and socialist countries that are still predominantly rural.

It is, perhaps, not surprising that in view of the levels and density of rural populations, and in the light of colonial and post-colonial emphasis on primary exports, that rural production levels in Pacific Asia are high in comparison to the rest of the Third World and continue to rise at a rapid rate, despite the shift of population into the urban areas. As Figure 4.2 implies, there is a distinction to be made between food and non-food production, but within the former there is another distinction to be made between subsistence and commercial food production. It is an unfortunate fact of life that many agriculturally rich countries produce primarily for export, so that many poor rural dwellers still suffer undernutrition and poverty in the midst of plenty.

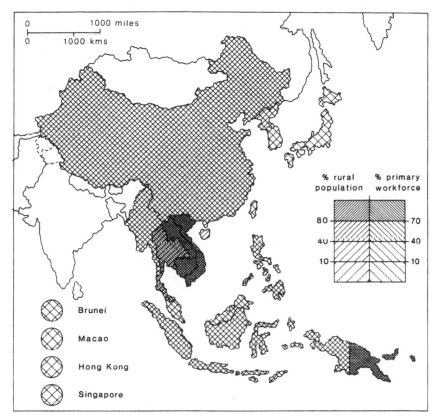

Figure 4.1 Pacific Asia: rural populations and primary workforce

Indeed, one of the most disturbing features of rural life in Pacific Asia is that it is still characterized by considerable underdevelopment compared to urban areas. It is not difficult to produce data to emphasize this contrast (Table 4.1), although we must be wary of interpreting such data as evidence of overwhelming urban bias in development programmes. One caution is that the rural and urban data themselves, if disaggregated, would show considerable variation in wealth and access to facilities. There are disparities within both urban and rural areas, and in general the poor as a group (whether urban or rural) are disadvantaged by development. This is brought out very clearly in the discussion on ethnicity, class and development in Malaysia in Chapter 6.

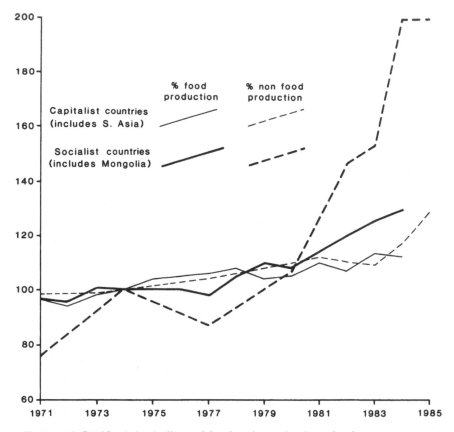

Figure 4.2 Pacific Asia: indices of food and non-food production

The second factor of which we need to be aware in interpreting indicators of disparity is the extent to which they reflect regional rather than sectoral contrasts. In other words, is the principal contrast between rural/agricultural and urban/industrial development or between a developed core (urban and rural) and an underdeveloped periphery? The distinction between the two is important not only in terms of identifying uneven development but also in explaining it and initiating policies to deal with it. As the discussion below indicates, many national governments do not fully appreciate the distinction between rural and regional (sectoral and spatial) disparity. As a result they introduce measures which attempt to combat one problem and yet worsen the other. Thailand, in particular, has adopted policies to combat rural poverty that have exacerbated regional imbalance.

Table 4.1 Pacific Asia: rural–urban development indices

| | % pop. access to safe drinking water | | % pop. access to safe sanitation services | |
	Urban	Rural	Urban	Rural
Brunei	100	95	75	28
Burma	36	21	34	15
Indonesia	40	29	31	30
S. Korea	88	60	100	100
Laos	28	20	13	4
Malaysia	87	71	100	59
PNG	55	10	91	3
Philippines	53	55	75	47
Thailand	50	70	50	44
Vietnam		31		70

Source: World Development Report 1990

There is a danger, however, in oversimplifying the dilemma facing governments in Pacific Asia into a set of simple dualisms – urban versus rural development, subsistence versus commercial agriculture. The real-world situation is much more complex than this. Most family farms have both subsistence and commercial production; whilst many of the apparently labour-abundant rural areas of Pacific Asia suffer severe labour shortages at crucial times of the farming year and are heavily reliant on recalling urban migrants to help out. At the same time the dearth of non-farm development opportunities in the smaller regional centres, together with the improvement of transport and communications, has meant that they have been bypassed in the migration of rural dwellers in search of new opportunities.

In short, the problem of rural development in Pacific Asia is not just a matter of increasing agricultural production. Disparity takes many forms, social and political as well as economic; it can affect both rural and urban populations; and it can have geographical as well as sectoral dimensions. But before we examine the various strategies for rural development and investigate what has happened in various parts of Pacific Asia, it is necessary to review some of the principal causes of the present unevenness in development in the region.

Explaining spatial inequality

As one might expect, much of the explanation for geographically uneven development can be attributed to colonialism. From the mid-nineteenth century onwards, an economic 'dualism' was introduced which favoured large-scale, commercial activities (such as mining or plantations) at the expense of small-scale, subsistence agriculture or rural handicrafts. It was not so much that the latter were deliberately disadvantaged, rather they were simply ignored. Thus land reforms or registration usually favoured the larger owners and dispossessed smaller tenants or group owners of land; the introduction of urban-based manufactured goods ruined local craft industries; traditional family values were eroded by exposure to western ideas; whilst gradual mechanization displaced rural labour.

In the post-independence period the situation did not improve to any noticeable extent. Planning strategies, whether capitalist or communist, have been transferred from outside the region and have emphasized urban industrial growth and the gradual trickle-down of benefits to the poor and to the rural areas. Within such development strategies, rural investment tended to result from sectoral programmes in education, health care or family planning, the goals of which were national in scope and were set by urban-based planners. This is a top-down approach, often assisted by foreign advisers and 'experts', which frequently fails to see the local perspective. Thus fertility control may be seen as desirable by the central government but is viewed as a loss of free labour by the peasant household; similarly, a new highway link to the capital city may simply extend commercial farming opportunities and also facilitate out-migration.

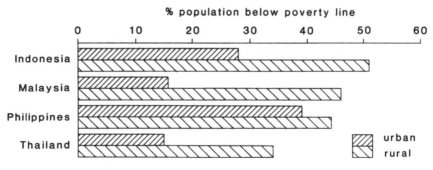

Figure 4.3 Southeast Asia: population percentage below poverty line

The consequence was development programmes that ignored and disfavoured peripheral areas and groups, both rural and urban, especially the poor (Figure 4.3). As a result, rural out-migration has been an ever-escalating problem throughout Pacific Asia, one which has resulted in immense problems of economic and social provision in the region's cities. In addition, changing values, such as the erosion of traditional dietary patterns and preferences, has resulted in a massive increase in food imports, even in countries with substantial rural labour forces and agricultural exports. In Papua New Guinea, for example food imports amount to 20 per cent of total imports by value and yet the country gains one-third of its export revenues from agricultural products. Clearly this is a trade-off which does not particularly advantage the bulk of Papua New Guinea's population, 87 per cent of whom are rural and whose average daily calorie intake is only 82 per cent of minimum requirements (see Case study G).

Previous development strategies have, therefore, fallen short of spreading the benefits of growth fairly throughout society and space. What were the responses to these shortcomings as far as rural and regional development were concerned?

Case study G

Food dependency

Dependency on food imports is one of the worst types of dependency because it can result in traditional food production declining to the extent that it cannot be re-established. In addition, food is an important basis of culture, and this can lead to the replacement of traditional cultural activities and lifestyles by imported ones; for example, eating less nutritious fast foods rather than home-prepared meals. Tourism reinforces these tendencies.

Food dependency can also lead to imbalanced trade and as prices rise this results in the need to export more to pay for the imported food. Ironically this can result in the export of commercial crops demanded by the developed countries, like palm oil, which further reduces the land available for basic foods.

What can be done to reduce dependency and increase self-sufficiency in food production? This not only means education of farmers so that they can produce more, but education of consumers into appreciating the value of traditional foods compared to less nutritious, imported ones. Such a reversal of preference

Case study G *(continued)*

trends will also demand an examination of media subversion of traditional values by its exaggeration of the virtues of modernization along western lines. This ranges from the saturation advertising of Coca-Cola in areas where the real thing (coconut milk) abounds, to the disastrous substitution of reconstituted powdered milk for breast-feeding.

This last point indicates that there is more than just food involved in this kind of dependency. It is symptomatic of a wholesale replacement of traditional values, many of which are valuable and wholly appropriate for Pacific Asian peoples, by western values which can undermine national economies and subvert lifestyles.

Policies for rural and regional development

There were, of course, always those who believed that the development strategies of the 1950s and 1960s were right and that in the fullness of time, with mature capitalist development, the benefits of growth would be felt throughout each developing country. A more positive attempt to devise a development strategy aimed specifically at the rural poor has been the so-called 'green revolution'. This topic is covered in depth in Chris Dixon's book in this series (*Rural Development in the Third World*) and I will simply provide an overview of its main features in so far as these have counteracted rural and regional disparities.

The initial impetus behind the development of new seed varieties of rice and wheat was to boost production of *basic* food commodities. This may seem to be what was required but many national governments did not interpret this potential rise in production solely as an improvement in household nutrition or incomes. Rather their objectives were national in scope. First, an increase in national production of food staples could either reduce import costs or even increase export earnings. Second, increased national production would enable urban food prices to be held stable and so reduce the pressure for wage increases in the city. Any benefits to individual farmers were sometimes of secondary interest.

As many observers have pointed out, the benefits in the rural areas from the new high-yielding varieties (HYV) of wheat and rice have been very selective in distribution, both spatially and socially. In some parts of Pacific Asia, wheat and rice were not the major food staples and so adoption was very limited; in other areas successful adoption of the HYVs depended upon parallel technological investments in insecticides, fertilizers or irrigation. Above all, adoption of the new systems (HYV and technology) was expensive and needed government subsidies. The necessary credit advances, educational instruction and other subsidies have been, inevitably, made available to the better credit risks, i.e. the larger farmers. For the smaller farmer, low-interest government loans had to be replaced by high-interest moneylender sources. For these small farmers the green revolution increased the risks from crop losses (HYVs are more susceptible to diseases, etc.). This is unacceptable to a poor family since crop failure means not just a loss of profit but starvation.

The new HYVs were part of a plant aristocracy and had a parallel social impact, benefiting the larger farmer and the already wealthy (see Case study H). In few countries, therefore, did the green revolution promote increased social and spatial equality; even when small farmers were involved, the increased production lowered market prices for wheat and rice. As far as most national governments were concerned, however, the green revolution was a success, apparently promoting rural/regional economic growth and favourably affecting their balance of payments.

In recent years the debate has been reopened with some observers claiming that 'technologically' the green revolution can bring benefits for all, citing as evidence the fact that in most Asian states few now starve in years of poor harvests. It is claimed that the adverse effects were more due to parrallel social and economic changes resultant from broader global processes.

The apparent failure of the green revolution to reach the poorest rural households has prompted a variety of responses from more sympathetic development strategists. The majority have been linked to what might be termed integrated rural development programmes which are discussed below, but an increasing minority have sought to re-emphasize traditional values in rural development planning. These range from encouraging the growth of traditional crops, which are often more suitable for local soils, more nutritious and reduce imports, to the resuscitation and expansion of small-scale cultivation, such as urban gardens.

Case study H

Malaysia: winners and losers in the green revolution

The Muda River area is located in the traditional rice bowl of
Malaysia and in 1970 a new irrigation scheme was initiated that
could permit double-cropping and the use of fertilizers in a
controlled environment. The World Bank provided half of the
funds for a scheme that had the twin objectives of increasing
Malaysia's self-sufficiency in rice and increasing farmers' incomes
on an equitable basis.

The immediate result was a hugh increase in the production of
rice, but was followed by an equally rapid levelling off. This
contributed substantially to overall domestic production and helped
decrease the dependence on rice imports.

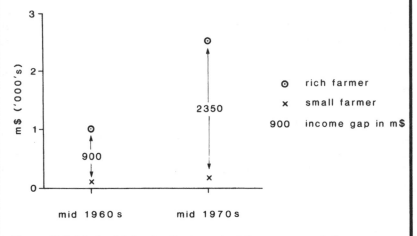

Figure H.1 Muda, Malaysia: the impact of the green revolution on
different income groups

The second objective, an equitable increase of incomes, has
been less successful. Incomes increased unevenly in the 1970s (see
Figure H.1) favouring those who were already better off and who
were able to take advantages of government assistance. As a result
the gap between the 'rich' and poor farmer widened during this
initial period of expansion.

Case study H *(continued)*

In the second half of the 1970s, yields remained fairly static, prices remained level (at best), but the cost of farm imports and consumer goods rose. The result was that all farmers experienced a real decline in incomes through their decline in purchasing power. The reaction of the Malaysian government was to subsidize fertilizer costs and raise the subsidy within rice prices. As the larger farmers use more fertilizer and produce more rice for sale, they have benefited disproportionately, thus widening the income gap further.

The real beneficiaries of the green revolution in the Muda Basin have been those who live in the nearby urban centres. Not only have they seen the price of rice deflate in real terms, due to increased production and subsidies, but they have also supplied the farm inputs and consumer goods that the Muda farmers have demanded. In effect, the regional and sectoral subsidy to rice farmers in the Muda region has been exported out to pre-existing social (rich merchants) and spatial (urban) clusters of privilege and wealth. This clearly illustrates the danger of an unintegrated rural development programme.

Integrated rural development

As David Lea and D. P. Chaudhri have shrewdly observed:

A careful look at the rural economy of most of the developing countries by a western-trained social scientist usually suggests that almost everything is wrong. Disease is widespread, health services poor, agricultural output low, roads few, merchants and money-lenders are exploiting the farmers, land ownership is skewed, administration is unimaginative and usually colonial in style. . . . It is obvious that all these things are somehow interconnected and there is a need to tackle them simultaneously. Thus the phrase 'integrated rural development' was coined.

It is important to realize that integrated rural development is not a radical development strategy. Indeed, it tends to mean all things to all people, ranging from integrating rural and national development (which was often the sole characteristic of previous planning), through integrating rural and urban development, to the integration of development within specific areas.

There has, therefore, been a recognition that rural development means more than the expansion of agricultural output and incorporates the following goals, according to Lea and Chaudhri:

(a) To improve the income and quality of life of rural residents
(b) To make rural areas more productive
(c) To ensure that development is self-sustaining
(d) To ensure local autonomy, usually through administrative and political decentralization.

Not all integrated rural development strategies place equal emphasis on each of these objectives and, moreover, a variety of methods has been used to promote whatever goals have been adopted. Some governments have opted for a westernized, technological approach, others have sought to promote socialist principles; some planning strategies have made great attempts to advance local participation in decision-making, others are still top-down documents. The result is a range of integrated rural development strategies with a variety of means by which anticipated objectives are to. be met. These are, of course, theoretical models. In reality there is a great deal of overlap in goals or methods. We will proceed to examine three countries which will illustrate these strategies in action.

The technological capitalism of the green revolution and its limitations have already been discussed above and also in Case study H. Thailand, our first focus, will furnish an opportunity to examine one rural economy where free-market forces have been half-heartedly fused with some new technological inputs with little noticeable success. South Korea and Vietnam, on the other hand, provide illustrations of successful rural development, ostensibly through reformist and socialist strategies. However, many of the means used to achieve success are remarkably similar despite their contrasting ideologies.

Thailand: regional imbalance and rural poverty

Historical geography

In physical terms Thailand comprises a central alluvial lowland surrounded on three sides by hills and plateau (Figure 4.4). Historically the Thai people originated from the Yunnan region of China and moved into what is now Southeast Asia many thousands of years ago, scattering around the region so much that Thai-speaking peoples are found in Burma, Laos, Cambodia, Vietnam and China.

These areas were always peripheral to the heart of the Thai state which, then as now, centred on the rich agricultural lowlands of the Chao Phrya river. Allegiances became severely strained when the British and French began to push in from their colonial territories in Burma and Indo-China in the nineteenth century, although both European powers saw Siam (as it then was) as a useful buffer area to minimize expensive direct colonial conflict. A series of treaties in 1855, 1896 and 1904 guaranteed the country's boundaries and left it with a degree of political independence. However, trade and commerce were very much in the hands of Europeans, notably the British and their ever-willing partners, the overseas Chinese.

By 1890, 70 per cent of the overseas trade of Thailand was handled by British firms, whilst internal trade was controlled by Chinese merchants. Furthermore, these expatriate interests influenced the pattern of economic growth to the extent that the country's former wide range of exports was drastically reduced until it was dominated by rice. This supplied the growing workforce in neighbouring Southeast Asian colonies, thus drawing a nominally independent Thailand effectively into the colonial system.

The expansion of rice production in Thailand was primarily accomplished without a restructuring of the land tenure system but was made possible by a series of measures aimed at centralizing the political and administrative system of the country. The peripheral regions and their villages lost their autonomy as local leaders became bureaucrats and imported goods began to erode small-scale industries. All this was paralleled, indeed was made possible, by the expansion of the railways from the late nineteenth century onwards. As Figure 4.4 clearly reveals, the new lines of communication focused firmly on Bangkok which began to assume primate proportions.

Between 1860 and 1940 the population of Thailand increased from 5.5 million to 15 million. This was due not only to natural growth but also

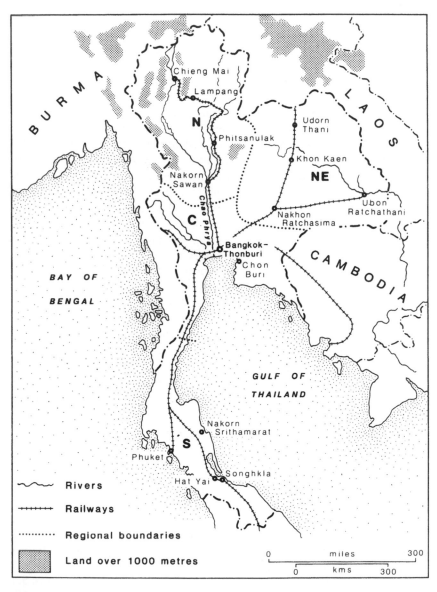

Figure 4.4 Thailand: map of the main regions

to immigration, particularly of Chinese who became the backbone of the modernization process, constructing canals and railroads, establishing rice mills and opening shops, whilst the Thai aristocracy concentrated on political and administrative matters.

The changed economic and social circumstances of the twentieth century, in particular the improved mobility (at least to and from Bangkok), have all resulted in massive changes in Thailand's regional balance. Essentially this involved the further subordination of the peripheral regions to a politically and economically dominant core centred on Bangkok. Some observers claim that regions no longer exist in Thailand which simply comprises metropolitan Bangkok and the countryside, but this appears to be an exaggeration and sharp regional contrasts can still be identified, although the urban–rural dimension is also a strong underlying theme.

Contemporary regional contrasts

Modern Thailand is undoubtedly dominated in economic and political terms by the central region, a fertile alluvial lowland set in the rain shadow of the Burmese mountains. Its natural richness, renewed each year by annual floods, has been improved and enhanced by irrigation and drainage systems that now network the region. Production is heavily dominated by rice but the adoption of green revolution technology has been slow owing to the Thai preference for their own types of rice. Nevertheless, production per capita has risen, although overall output has remained fairly stable.

This apparent contradiction is due to the loss of agricultural land each year to the vast sprawl of Bangkok. Originally sited on the low-lying land of the Chao Phrya River for defensive purposes, the capital has become the most extreme example of urban primacy, not only in Pacific Asia. With its growth accelerated in recent years by improved communications that have enabled migrants to move directly to the capital over much longer distances, Bangkok dominates Thailand in every conceivable way. This has resulted in a population of some 7 million, some 70 per cent of the urban population of Thailand, and makes the capital 50 times larger than the second city of the country – a discrepancy that has grown wider over the years. Indeed, only two other urban centres top 100,000.

Yet this dominance is achieved by a city housing only 15 per cent of Thailand's population and indicates the acute contrast that exists between capital and countryside. Moreover, if over 80 per cent of Thais still live in the rural areas there is clearly going to be some regional differences within the extensive area that lies outside metropolitan Bangkok and the central plain.

The smallest of these other regions is the South, which comprises about 12–13 per cent of the country's land and population. It is mainly equatorial woodland but produces increasing amounts of rubber, tin and rice. The South also has a distinct Malay and Muslim character. In contrast, the northern region is an area of deeply dissected and heavily wooded hills. It has good agricultural areas in the valleys, and its pleasant climate and beautiful scenery have spawned a burgeoning tourist industry centred on the regional capital of Chieng Mai.

The final region is the poorest in Thailand and will be the focus of this investigation. The Northeast constitutes about one-third of the country's land surface and population. It is made up largely of the sandstone Korat plateau which is drained by the Mekong River and its tributaries but which tends to be marshy and boggy in summer, dry and barren in winter. Most agriculture is subsistence in nature, centring on maize rather than rice, and the region contributes only 20 per cent of the national GDP.

The Northeast has always been the poorest of Thailand's major regions. The large Lao minority and its isolation (the railway to Khon Kaen was not completed until 1935) has meant that central controls were limited, banditry rife and when the first Thai family income survey was taken in 1929, the northeastern average was only 30 per cent of that of the central plain. Such poverty of income and production persisted through the first half of the century, with little response from the national government, until the 1960s when the escalating conflict in Vietnam, Laos and Cambodia began to affect the stability of the region.

The Northeast had always been prone to lawlessness and dissatisfaction but since the 1960s this has been labelled 'communist inspired'. The result was increased attention from the government, prompted by United States assistance towards the construction of military bases in the region. The roads that accompanied this served only to accelerate the already high out-migration, particularly to Bangkok to which the regional capital is now linked by a regular, direct bus service.

What was the government's response to this deepening rural poverty and regional imbalance?

Regional and rural planning policy in Thailand

It must be noted at the outset that regional growth strategies in Thailand have been motivated not by a concern for rural and regional inequality but by the growing political instability of the border regions and by the escalating problems of Bangkok. There are several national growth strategies that affect regional development in Thailand. First, there has been a desire to diversify both within agriculture and also away from the primary production that dominates employment and exports. The government has thus given every encouragement to the growth of manufacturing, whether domestic or foreign financed. To be fair, it has had a substantial amount of success, but increasing commercial agriculture (such as pineapple plantations) serves to displace small farmers, whilst further investment in urban industrial expansion encourages out-migration to Bangkok. These have worsened regional disparity. Indeed, although manufacturing (20 per cent) now contributes more to GDP than agriculture (19 per cent), it reflects underproduction in the rural areas as much as the undoubted expansion of export industries. This is an important factor because sustained manufacturing growth will depend as much on the expansion of Thailand's domestic market (by improving incomes and purchasing power) as on export markets.

Second, the Thai government interprets poverty in sectoral rather than spatial or regional terms. Thus low incomes are seen as being associated with the agricultural sector rather than certain parts of the country. The response is to induce a shift from agriculture to industry which is, of course, urban based. The Thai planners have tried to decentralize this sectoral emphasis on urban economic growth by means of a growth-pole strategy, particularly in the principal regional cities of Hat Yai/Songhkla, Chieng Mai and Khon Kaen. Promoting regional universities, ports and tourism has not, however, had much regional or rural impact.

One cannot escape the impression that Thailand has no coherent regional or rural development strategies, only national development programmes that impinge on regional or rural development, usually adversely. Indeed, the only positive policy that the government has been seen to promote in the regions has been that of birth control.

The consequence for the northeast throughout the 1960s and 1970s was a continued worsening of its position relative to Bangkok and the central region (Figure 4.5). Not surprisingly, as the regional economy stagnated, the propensity to migrate rose: by the late 1970s, 43 per cent of Bangkok's migrants were from the Northeast. In many villages up to three-quarters of the households had experienced migrant loss at some time or other. Interestingly, the improved transport has not only made it easier for villages in the Northeast to migrate to Bangkok, it has also

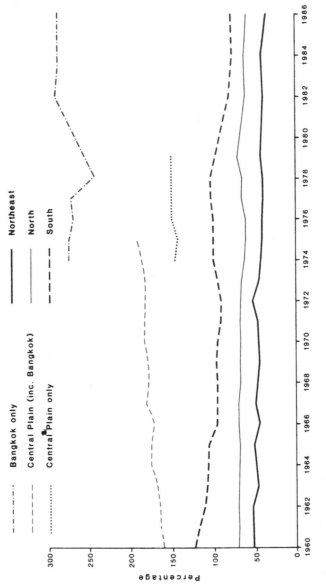

Figure 4.5 Thailand: regional contributions (%) to national GDP

made it easier for them to return. About half of the migrants, therefore, are resident in the capital for less than three months before they return home. However, this pattern may be repeated many times.

The failure of government policy to address regional imbalance led in the 1980s to external advisers recommending a 'basic needs' type of approach. Aid funds have therefore sought to fuse the technological approach of the green revolution with the reformist strategies of a fairer distribution of growth within small-scale projects that aim to sacrifice growth maximization for a more equitable development programme. Unfortunately this has proved to be somewhat rhetorical as there is little growth in the northeast to sacrifice. Moreover, locally prominent farmers have tended to monopolize the benefits of growth in the absence of any state interest in supervisory controls to ensure equity. The result for many farmers, given media encouragement of rising expectations, has been increased resentment, dissatisfaction and instability.

Vietnam: reunification, rural and regional development

Historical geography

Vietnam, like Thailand, is still a predominantly agricultural economy with more than 80 per cent of its 62 million people living in rural areas. Nevertheless, the country is mountainous and only 5 per cent of the area is intensively farmed, particularly the river deltas of the Mekong in the south and the Song Koi (Red) river in the north where high population densities are found (Figure 4.6). As noted in Case study B, after hundreds of years of independence the Vietnamese empire was absorbed by the French in the late nineteenth century.

The distribution of natural resources suggested that, as during the colonial period, the south should produce food surpluses whilst the north, with its minerals, coal and labour supply, should become the industrial heart. However, the partition of the country in 1954 into a communist north and a capitalist south, together with the ensuing war, totally distorted this economic scenario. North Vietnam had to become self-sufficient in food, which it did by means of integrated rural development and collective agriculture. On the other hand, the demands of the military and the incessant bombing of its towns meant that its urban industries did not develop to any great extent.

In South Vietnam, in contrast, agricultural production fell as military and political control of rural areas was lost – imports from the United States substituted. At the same time rural refugees swelled the urban

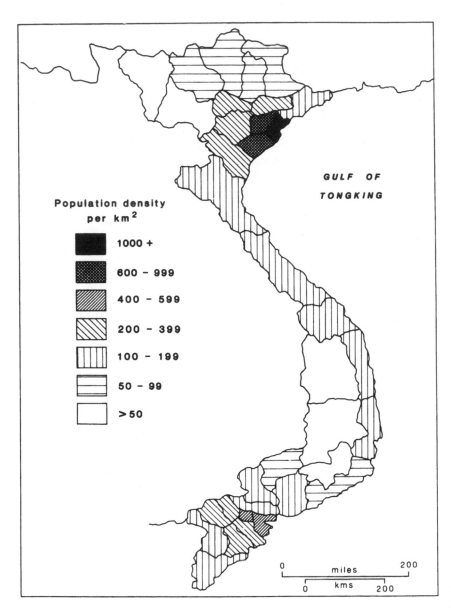

Figure 4.6 Vietnam: population densities

population, particularly in Saigon, leading to the development of consumer and service industries. As a result, urban population proportions of total population were dramatically different on the eve of reunification in 1974. In the north it was only 11 per cent but in the south 65 per cent were urban.

As far as rural development is concerned, the north clearly developed a system through which it could increase production and redistribute wealth according to its socialist objectives. The question in 1975 was whether it could achieve the same results in the capitalist south. Could effective sectoral change be given a spatial extension?

Socialist rural development in the north to 1975

The immediate policy when the French left in the mid-1950s was land redistribution amongst the peasant and landless. But high population densities led to excessive fragmentation and low levels of production. By the late 1950s the government decided to encourage the develop ment of collectives based on villages: first, through cooperation within villages and then between villages. These larger production units proved to be more efficient and by 1965 two-thirds of peasant households were members of high-level communes; by 1975 this had risen to 93 per cent.

In the collectives, production occurs through brigades which earn workpoints. The equivalent of 150–200 days' work qualifies an individual to purchase food at subsidized prices and also to receive other benefits. The latter constitute an important aspect of rural development and have resulted in the rapid spread of literacy and health care programmes in rural areas. These successes are not just the result of the adoption of the collective system but also attest to the fact that unlike most capitalist states, the rural areas have a much higher priority than urban areas in the distribution of goods and services.

Rural change in the south after 1975

The 'traditional' method of farming in the south was fundamentally different to that in North Vietnam. The restructuring by the French had resulted in a region of large landowners and many tenant farmers, each tending to be scattered thinly across the region. There was no long-established village commune to act as a basis for collectivization. More-over, during the 1960s the farmers had come to be heavily reliant on the central government for the technological inputs, particularly fertilizers, to a green revolution aimed at increasing production to feed the burgeoning cities.

The result was that early attempts to collectivize the south met with little success. Even the widespread 'rustication' of thousands of urban

dwellers into the countryside as farm labour could not help maintain productivity since agricultural production was based on technological rather
than labour inputs. As a result the government, which desperately needs
the food, has been forced to recognize the continued existence of private
production and continues to supply fertilizers to the farmers (Table 4.2)
whose dominant links are therefore with the state and not other farmers.

Table 4.2 Vietnam, South Korea and Thailand: comparative rural development
indicators

	Vietnam	South Korea	Thailand
Cereal imports (k tonnes)			
1974	1,854	2,679	97
1987	653	8,758	255
Food aid (k tonnes)			
1974	6	234	9
1987	76	–	18
Index food production p.c.			
1974–81	100	100	100
1985–87	114	100	107
Fertilizer imports (kg/ha)			
1970	512	2,466	76
1986	620	3,853	236
Cereal production (kg/ha)			
[% change since 1974/5]	2,604 [22]	5,498 [33]	2,085 [11]
% Distribution of agricultural land/size holding (1980) Vietnam (1960)			
under 5 hectares	61	100	39
5–50 hectares	29	0	61
over 50 hectares	11	0	0
% rural pop. with access to			
safe drinking water	31	70	60
sanitation services	70	44	100
Av. % change in population (1975–86)			
rural	2.3	−2.2	2.1
urban	2.9	4.9	3.4

Sources: World Development Report (1989), *World Resources* (1987)

Present and future trends

The Vietnamese government does not rely on agriculture for all of its
development planning. It wishes to diversify and develop its industrial
potential in the northern Red River delta and hopes to achieve this
objective through the amalgamation of rural collectives into much larger
agro-industrial collectives which will begin to meet their own immediate
industrial needs (for cement or fertilizers, for example).

Success, it must be said, has been patchy. Loyalties in the north are to local agricultural collectives which are based on traditional village communes. Reorientation of individual loyalties to larger regional entities has proved to be very difficult. One response by the state has been to use the capitalist initiatives which infiltrated north after reunification to increase production outside the collective system through the exploitation of 'slack-time', or time not spent on work for the collective.

Slack-time is used either to encourage private production on small private plots of land, both for the family and for sale to the state, or to develop agro-industrial production via subcontracting to the home simple work from state factories, such as jute-stripping. These measures have proved popular but it is a matter for debate as to whether they point the way to the future or constitute a temporary measure whilst the full incorporation of the south and the proper organization of agro-industrial collectives occur.

Although the Vietnamese government has clearly been successful in lessening dependence on imports (see Table 4.2) and in redistributing the benefits of growth more evenly in both spatial and social terms, we must not be overly optimistic about the present state of the country. Redistribution of growth means little in real terms if growth is minimal, and in Vietnam this is the case. It must be remembered that 30 years of continuous war led to massive destruction of human, capital and environmental resources, from which Vietnam has struggled to recover. It continues to be a very poor country and yet, because of the threat from China and its own ten-year occupation of Cambodia (for strategic and humanitarian reasons), its military budget has been high. Poor harvests in the mid-1980s added to the country's burden. But significantly these natural disasters provoked widespread complaints but not widespread starvation. Currently Vietnam is a growing rice exporter and prospects for the future look positive.

South Korea: growth and equity

Historical geography

The Korean peasantry had a continuous history of exploitation within a feudal type of system from the fourteenth century to the 1970s. Direct feudalism lasted until 1910 when Korea became a Japanese protectorate but, although colonialism resulted in more investment in the rural areas, most of the produce and profits went to Japan and rural indebtedness increased. Within this quasi-feudal system, 90 per cent of the peasantry lived on tiny tenant farms, with only about 2 per cent of the households owning more than half of the land, and throughout the present century increasing numbers drifted to the new cities to work in Japanese factories.

The end of Japanese control resulted in a very similar situation to that of Vietnam. A struggle between capitalist and communist factions with substantial external intervention began in 1950 and ended in 1953 with a division of the country between a communist, more industrialized north and a capitalist, agricultural south.

Immediately after the Second World War, when the country was under United States military control, the government tried to reform land ownership but entrenched interests effectively prevented this until the total disruption of the Korean civil war provided an opportunity to start afresh. In the reorganization of the 1950s, maximum holdings were set at 3 hectares per family. However, the immediate impact was social more than economic; a final destruction of feudalism and its related social structures. The new pattern of small holdings remained intact with little amalgamation until the 1970s, when 94 per cent of the cultivated area was still farmed in plots of less than 3 hectares, by which time new rural forces for change were being felt.

The late 1950s and 1960s were, therefore, periods of rural stagnation and the new US-backed government of Park Chung Hee which assumed control in 1961 had, by the time of his assassination in 1979, witnessed an economic transformation. During the 1960s this was almost totally based on urban industrial expansion which subsequently provided the funds for rural development.

Urban industrial growth in the 1960s: the prelude to rural change

The rapid expansion of export-industrialization in South Korea is discussed in Chapter 7. At this point it is sufficient to note that it relied very heavily on United States and Japanese aid and trade, being seen as an important front-line capitalist state in the economic and political struggle against communism. The continuance of the war ethos and the traditional discipline of Confucianism both permitted the development of a strong central control over the economy. This was state capitalism not *laissez-faire* capitalism.

The massive economic expansion of the 1960s, in which growth averaged 18 per cent per annum, had several repercussions in the rural areas. First, it resulted in the neglect of agriculture. There was little investment, minimal mechanization, and few non-farm opportunities in the small regional towns. Unemployment and underemployment grew rapidly in rural areas where per capita income was falling far behind that of urban workers (Figure 4.7). The result was a huge labour shift from the countryside into the cities.

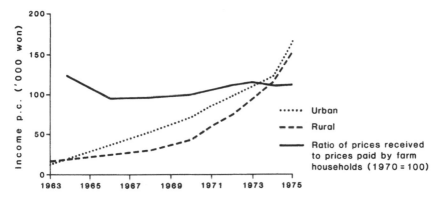

Figure 4.7 South Korea: urban and rural incomes and terms of trade

The corollary of this rural poverty and population migration was accelerating urban problems, particularly in Seoul which grew at 9 per cent per annum through the 1960s and absorbed three-quarters of all the natural population increase of the country. The government response was to deflect this expansion to other urban growth poles, but for the most part this affected only Pusan, so that between 1970 and 1975 over 90 per cent of the country's population growth occurred in the two main cities. In contrast, between 1960 and 1975 the population in the rural townships (of less than 50,000) fell from 36 per cent to 19 per cent of the total population. Clearly something had to be done for the rural areas to forestall a crisis.

Rural change since the 1970s: the new village movement

The immediate spur to a restructured rural policy was the realization that agriculture was not producing enough food for the burgeoning urban (and military) populations. It was decided that this was because of the low prices paid for farm products; consequently the government introduced a subsidy which gave the farmers a better income but still kept prices low for the urban consumers. The result was a noticeable improvement in the terms of trade, i.e. the comparison between income and expenditure, for rural households (see Figure 4.7). The government also began to encourage technological change along the lines of the green revolution and by the late 1970s South Korea had the highest paddy rice yields in the world, but it imported vast amounts of fertilizer (see Table 4.2). Mechanization was, however, difficult on farms of such small size.

What made these policies different from those in other Pacific Asian countries was the context in which they were introduced to the rural areas. The measures formed part of the Samael Undong or new village programme which began in 1971 and is basically structured around the village level cooperative but retaining private land ownership. The villages themselves select development projects and carry them out with government financial and professional assistance. In short, there is a substantial degree of 'planning from below'. The difference from Vietnamese collectives lies in the extent of private ownership and the incentive to increase rural incomes (Figure 4.7) and the quality of rural life.

The programme was organized in a series of stages of cooperation with the earlier phases concentrating on infrastructural improvements and later moving on to agricultural productivity. Undoubtedly the material achievements have been substantial, with both rural incomes and the quality of life improving dramatically since the mid-1970s. Moreover, there is genuine cooperation not only between peasant households but between the peasantry and the state. For the first time ever, a fair and efficient rural bureaucracy has emerged, largely because many of its members have been recruited from the peasantry, itself a consequence of the improved levels of rural education.

And yet the new village movement has not been without its critics. The improved incomes quoted for households are substantially moderated by the continued large size of rural households, many of whom continue to be underemployed so that rural income *per worker* is still less than half of that of the city dwellers. Moreover, although the new village movement has brought spatial improvements in one sense (rural *vis-à-vis* urban), its regional impact has been more distorted with the farms in and around the main development corridor between Seoul and Pusan obtaining maximum assistance from price subsidies. In contrast, rural households elsewhere, which are more subsistence- rather than sales-oriented, have received fewer benefits.

The principal doubts about the new village policy are, however, not so much economic as political. Critics point to the contradiction of political liberalization in rural areas with a steadily increasing repression at the national scale. Some Asian governments feel that the new village movement has generated instability by raising hopes of genuine democracy. Certainly the frequent urban disturbances in South Korea seem to support this contention.

This instability and uncertainty beyond the level of the cooperative have made it difficult for the government to move to larger-scale units of rural and regional productivity. As in Vietnam, loyalties remain primarily to the village and there is suspicion of broader-based enterprises. As a result it has been difficult for the rural development programme to move beyond infrastructural and agricultural projects into areas of non-farm employment. Indeed, employment in small-scale industrial activity has steadily declined and three-quarters of all manufacturing employment is now concentrated in Seoul and Pusan. Massive corruption and appropriation of funds in the central government offices has not helped this situation.

But perhaps the biggest problem is that rural development has clearly been subsidized by, and is dependent on, continued success in export industries. Growth in incomes has come from state subsidies rather than increases in productive capacity. Table 4.2, therefore, reveals that food production indices are static, despite structural change and technological support, whilst cereal imports have more than tripled since the mid-1970s.

The result is a somewhat brittle and fragile rural situation. If export earnings fall, as they did during the mid-1980s, rural development will suffer and voices of discontent will be added to the constant urban clamour for political reform.

Conclusions: a comparison of Thailand, Vietnam and South Korea

It is difficult to sit in judgement on the successes and failings of rural and regional development programmes in these three countries. Different perspectives give rise to quite varied interpretations. What is seen as successful by national planners is often evaluated very differently by the farmers themselves. What we can do is to highlight the common denominators in terms of rural development within the three countries and try to anticipate the nature of problems facing future planning.

First, it is evident that Vietnam and South Korea have strong central government control of their rural areas, which is not the case in Thailand despite the enormous extent of urban primacy. Second, South Korea and Vietnam have made the village the focus of their development programmes, building on an unbroken tradition of community loyalty and obligation at this level. Thailand, in contrast, has an institutional–individual structure for its rural development, with village loyalties being deliberately weakened as part of its modernization process. Again, the achievements of South Korea and Vietnam, unlike Thailand, were based on the twin pillars of infrastructural investment and land reform with the government underwriting these measures.

Despite their successes (from either top or bottom perspectives), both Vietnam and South Korea face similar problems. Cooperative ventures have proven difficult to escalate beyond the village level, particularly as this involves greater sacrifice of private or personal production units in favour of larger ones with which individuals cannot identify and therefore show loyalty. There have also been problems in diversifying away from agricultural activities to other forms of employment. All three countries have recognized this but have preferred to expand their industrial production in large cities rather than small towns. The result everywhere has been a shift of population to the urban areas, admittedly of varying intensities (Table 4.2). In many ways, therefore, rural development is closely related to what happens in the cities. In South Korea rural change has been subsidized by urban-industrial prosperity; in South Vietnam it has been at the expense of export-oriented investment; in Thailand rural resources, both human and physical, seem to have fuelled the overwhelming dominance of metropolitan Bangkok.

Clearly it is impossible to separate rural and regional development from other spatial and sectoral aspects of economic change. What is needed is a determination not only to reallocate some priority in development objectives to the rural areas, and to provide the means to achieve this, but also to ensure that rural households themselves, particularly the poor, are incorporated into the decision-making and administrative process.

Key ideas

1 Although there are rural–urban contrasts in standards of living, there are also many variations within rural areas.
2 The green revolution did not benefit the rural poor as much as other groups.
3 Current policies favour integrated rural development strategies.
4 Success or failure in rural development is not necessarily related to political structure.

5
Population growth and mobility

Demographic trends in Pacific Asia

Table 1.1 contains some of the basic demographic indicators for Pacific Asia and shows clearly the considerable gap which still exists in the region between birth and death rates. As with many parts of the Third World, the rapid diffusion of improvements in health care from developed countries has brought down death rates throughout the region, with one or two exceptions. Birth rates, on the other hand, which are influenced by a much more complex mix of medical, social and economic factors, have been brought down more slowly and rather erratically.

The result is that population growth rates in Pacific Asia, whilst being relatively lower for the region as a whole compared to the rest of the Third World, remain quite high in some individual countries. When added to the generally youthful population structure, this means that many countries will still be growing rapidly for some time to come (Figure 5.1).

Allan and Anne Findlay have discussed in detail elsewhere in this series the nature of demographic change and its relationship to the development process. It is sufficient to note at this point that low-income households and national governments tend to have very different views on the merits and demerits of large families. Most of the poor see large numbers of children, especially males, as sources of unpaid labour and as security in old age in countries without welfare systems.

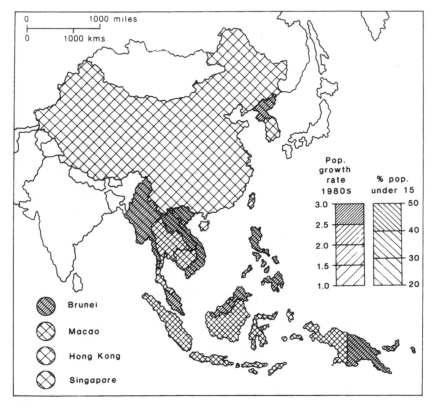

Figure 5.1 Pacific Asia: population growth rates and percentage under 15 years

Moreoever, cultural prejudice or lack of education mitigates against change.

In contrast, most governments see large families as detrimental to economic growth by putting pressure on limited resources, whether food, shelter or other kinds of basic needs, and by reducing the level of personal savings. During the 1960s many development strategists became very pessimistic about the rate at which population growth in the Third World was putting pressure on finite world resources. This neo-Malthusian panic led to recommendations for substantial curbs on fertility and the major world institutions, such as the World Bank and United Nations, have strongly supported these measures over the last two decades. Most governments in Pacific Asia have enthusiastically promoted family planning, particularly amongst women, and acceptance rates have changed dramatically in some countries. In Thailand, for example, the proportion of women of child bearing age who use contraception has risen from 15 per cent to 65 per cent since 1970; in Malaysia the change was from 7 per cent to 51 per cent, and in the Philippines from 2 per cent to 44 per cent.

However, birth rates do not come down solely by means of family planning campaigns. Acceptance of the arguments for smaller families is strongly related to the socio-economic status of the family which, in turn, is a reflection of revised family values and priorities, themselves a function in particular, of the level of female education. This suggests that reduction of birth rates to levels approaching those of Europe and North America will depend as much on broad socio-economic changes within society as a whole, as on incentives or disincentives related to fertility levels. Unfortunately in too many countries, family planning is seen as a substitute for development rather than simply one component of it. The lowest rates of population increase in Pacific Asia are therefore found where both economic growth and strongly supported family planning programmes co-exist, i.e. Hong Kong and Singapore. Ironically, Singapore and Malaysia have decided, for different reasons, that they have been too successful in slowing their population growth and both have reversed their population policies in recent years (see Case study I).

The obvious consequence of this continued delay in the demographic transition, i.e. the reduction of birth rates to the same extent as death rates, has been acute population pressure on resources and basic needs. In Pacific Asia this has particularly been felt in rural areas where the bulk of the region's population continues to live, where families continue to stay large, and where continued commercialization of agriculture has weakened traditional ties to, and income from, the land itself. As a consequence, many rural dwellers have reacted to these difficulties by moving to areas where opportunities are perceived to be better. Although the great majority of such moves are to towns and cities, this is by no means exclusively the case and before examining the details of the major types of population mobility in an Indonesian context, it may be worth reviewing the full range of movements that affect Pacific Asia.

Case study I

Singapore and Malaysia: family planning in reverse

Throughout the Third World over the last four decades, population policies have tried to curb family size and reduce pressure on limited national resources. In the 1980s, however, two Pacific Asian countries sought to reverse these policies but for quite different reasons.

Since establishing its republican status in 1965, Singapore has successfully curtailed population growth through a series of strong disincentives such as the withholding of maternity benefits for the third child, demotion to the end of housing and school waiting lists, and so on. Indeed, so successful was the 'stop at two' policy that, in conjunction with rising affluence and levels of education, the birth rate matched the mortality rate by 1975.

By the 1980s the government was becoming concerned about this success because it was jeopardizing Singapore's main economic resource – its labour force. Estimates suggested that the labour force would begin to decline by as early as 2010. As a result a series of new slogans and incentives encouraging larger families has begun to appear.

However, Singapore's economic growth is now based on hi-tech industries and the government is anxious to create an educated, skilled workforce. The result has been population policies which favour the educated. From the early 1980s the Singapore government has sought to encourage educated women to marry earlier and have more children. At one stage the Prime Minister even lamented the demise of polygamy as one solution to the dilemma. More pragmatically, quotas were placed on the number of women in education institutes. A Social Development Unit was also established to bring together graduates with a view to eventual marriage; free honeymoons at the Malaysian Club Med acted as additional incentives.

By 1987 this concern was institutionalized within a New Population Policy which contains a mixture of fiscal and social incentives to encourage three-child families. However, this encouragement is heavily biased toward the better educated parents. It is reported that women with at least five 'O' levels can claim up to S$10,000

Case study I *(continued)*

(about US$5,000) for each of their first three children; less educated women receive far less generous incentives.

Malaysia currently has a population of some 16 million, but its current development plan refers to a population target of 70 million by the end of the next century. The reasons for this enormous growth also seem, at first glance, to be economic. Malaysia, it is argued, needs a much bigger labour force to exploit its resources and to industrialize. However, there are signs that even the modest increase in the rate of population growth during the 1980s failed to be matched by the creation of jobs. Digging a little deeper reveals other, more disquieting, reasons for population growth.

Disaggregating current population growth figures reveal quite marked ethnic differences in the overall picture. The fertility rates for Malays are rising, but those of Chinese and Indians are falling. The average Malay family has four or five children, the average Chinese family has two or three. This contrast must be placed in the context of economic and political rivalry between the principal ethnic groups in Malaysia (see Chapter 6). Indigenous Malays have long felt that they have been overshadowed in their own country and some see the restoration of substantial numerical superiority as an important step in reversing the present position. (see Denis Dwyer (1987) for a fuller discussion)

Population movements in Pacific Asia

A wide range of population movements has affected and continues to affect the region. By and large those linking Pacific Asia to other parts of the world are historical, such as European movements related to colonization and decolonization, or the contracted labour shifts related to this. The legacy of such population shifts are clearly substantial and have already been discussed in Chapter 2. A more recent international movement has occurred over the last ten years with the rise of contracts awarded to Asian firms for construction projects in the Middle East. Increasingly, many of these firms are South Korean who transfer their own labour to the Gulf States where they reside perhaps for several years until contracts are completed. The impact of such labour migration on remittances and incomes in East Asia can be considerable.

Another form of labour represented in the Middle East is domestic labour, with Filipino and Thai women being dominant in this migrational flow. In total, there were some 700,000 workers from Pacific Asia in the Middle East in 1981.

The final type of international labour movement involving Pacific Asia is its 'brain drain'. Educated and trained workers have always shown an inclination to move to developed countries where salaries are higher. Even by 1980 some 250,000 Asians were entering the United States each year but in recent years this loss of human resources has become even more pronounced, particularly in Hong Kong in the run-up to the return of the colony to China in 1997. Whilst overall statistics are difficult to obtain, it is instructive to note that in 1988 the largest bank in Hong Kong lost 50 of its 700 executives through emigration overseas.

Similar movements of labour have occurred and continue to occur within Pacific Asia as a whole. One of the most extensive and long-lasting has been the emigration of Chinese from their home country to cities all around the region. These overseas or Nanyang Chinese, as they are usually known, continue to form the backbone of urban commerce throughout most of Pacific Asia. Integration with the host community has varied considerably, with ethnic tensions persisting in many parts of the region between indigenous and immigrant populations (see Chapter 6). Since most countries became independent in the 1940s and 1950s, freedom of movement for economic migrants from China has been far more constrained and only a comparative trickle are able to make their way into Hong Kong, although their impact on that small entity continues to be substantial (see Case study J).

Labour migration continued to occur throughout the region where economic growth was strong enough to attract people from other countries. For example, Koreans were allowed into labour-short Japan in the 1940s following decolonization. Later in the 1950s and 1960s Singapore willingly acted as a focal point for the migration of urban Chinese during the early xenophobic stages of independence within Southeast Asia when the Malay peoples of the region gave vent to a long-suppressed economic envy of Nanyang Chinese. Singapore continues to be a magnet for international labour, but this time for Malaysians and Indonesians seeking to take advantage of higher wages in the city. Such labour migration is, however, tightly controlled by Singapore. One final contemporary labour movement in the region is the movement of Filipino maids to the more affluent cities such as

Singapore and Hong Kong, where some 45,000 work in total. Along with other migrant workers, they remit some US$1.5 billion to the Philippines' economy each year.

Most of the intra-regional movement of population that affects Pacific Asia today is unfortunately that of the many political refugees. The largest and earliest of the post-independence movements of this nature was the flight of the Nationalist Kuomintang to Taiwan after defeat in the Chinese civil war in the late 1940s. Their settlement on this island has, of course, transformed its economy totally. Over the 40 years since this traumatic event, there have been continued but sporadic movements of refugees from other regional conflicts – usually involving struggles between conflicting political ideologies.

The most notable refugee movements have, of course, primarily involved Southeast Asian countries, with huge numbers of Khmer and Lao fleeing from genocide and invasions in the late 1970s and early 1980s. However, for the last ten years there has also been a steady, occasionally massive, outflow of refugees from Vietnam. It is estimated that since 1975 some one million refugee movements have occurred in Southeast Asia. Over half of these are Khmers now resident in refugee camps just inside the Thai border. Most of the remainder are Vietnamese boat refugees, an extensive number of whom are incarcerated in holding camps in Hong Kong.

Political turmoil can also create refugee movements within countries. During the Vietnam War, for example, the population of Saigon and other cities in the south virtually doubled as Viet Cong troops assumed control of rural areas. Once Vietnam was reunified the government introduced programmes of de-urbanization or rustication as 'surplus' urban residents were (unsuccessfully) decanted to new economic development zones. An even more drastic and catastrophic de-urbanization programme occurred in Cambodia under the anti-intellectual programme of the Khmer Rouge.

Within individual countries the most dominant form of population movement is to towns and cities which is discussed in some detail in Chapter 8. The response to this massive shift in population has varied enormously. Some governments have simply tried to stop the movement of people by legal means, requiring permits before allowing residence in the city; others have sought to improve economic and social conditions in the rural areas in order to encourage more people to stay. However, irrespective of the incentives or disincentives employed, the inexorable shift of population to the cities has continued. We will examine in more detail below the nature of this movement in an Indonesian context.

But finally to complete this broad overview of population movements in Pacific Asia, it must also be pointed out that not all rural out-migration is to urban areas. Many countries of Pacific Asia still have rural lands to develop. Land clearance and settlement programmes have been strongly supported for a number of years in the region, most extensively in Indonesia and Malaysia. Again, a closer examination of one of the schemes follows.

Case study J

Hong Kong: urban or international migration

To many planners Hong Kong is a unique city, sitting in its own backyard of 1,000 square kilometres and devoid of all the usual pressures resultant from rural migration. This, of course, is far from accurate and Hong Kong could not have reached its present population of 5.5 million nor its economic prosperity without in-migration. In other circumstances such migrants would be considered part of a normal rural-to-urban migration flow, but the existence of a frontier makes this movement international. This has enabled Hong Kong, consciously or not, to screen its migrants to a certain extent and prevent the build up of 'excessive' numbers.

The waves of migration from China into Hong Kong have always ebbed and flowed, at least until 1948. Usually migration out of China was triggered by internal instability, a situation which occurred with increasing frequency from the mid-nineteenth century onwards. Thus the Taiping and Boxer rebellions were periods of rapid growth, as were the 1911 revolution and the bitter Sino-Japanese war.

During the Japanese occupation, the population of Hong Kong fell from 1.8 million to 650,000 but regained its former population by 1949. There followed a huge influx of refugees from the civil war on the mainland and although, after the communists won power, many returned to China, enormous numbers also settled in Hong Kong.

As migration grew in scale, so did the catchment area of the migrants themselves. By the 1930s and 1940s, the local flow of

Case study J *(continued)*

economic opportunists from the Pearl River region had been overtaken by political refugees from all over China. The latter proved to be of fundamental importance to Hong Kong since many were urbanites, from cities such as Shanghai, who brought with them capital and entrepreneurial or manual skills. These helped transform Hong Kong's economy in the 1950s from a trading port to a manufacturing centre. Although the Chinese border was officially closed in the 1950s, there has been a steady flow of illegal and legal refugees, averaging some 40,000 per annum during the 1950s and 10,000 during the 1960s. These have been very important in helping keep down labour costs in the colony.

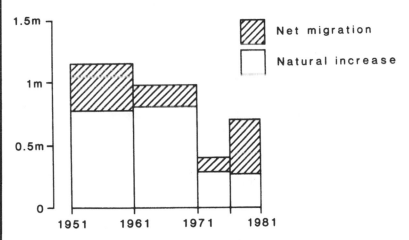

Figure J.1 Hong Kong: population growth components

Apart from one or two border relaxations, the immigrant element within Hong Kong's population fell steadily in the post-war years, being surpassed by natural increase for the first time, even though the birth rate was also falling (Figure J.1). How-ever, this changed suddenly in the late 1970s when immigration

Case study J *(continued)*

suddenly rose once more, both from China and Vietnam, and almost half a million arrived in Hong Kong by the early 1980s. This was partly curtailed by the rescinding of the very British 'touch base' policy towards illegal immigration, whereby those who managed to elude the army and police patrols and 'touch base' in Hong Kong itself, could apply to stay. The sheer volume of successful home runs eventually forced the authorities to abandon this jolly sporting practice.

Most migrants exhibit traits not too dissimilar from those of normal rural-to-urban population shifts. The legal migrants tend to be dependents joining accepted migrants, whilst the illegals are usually young men (70 per cent aged between 18 and 24) in search of better paid work. Of the illegal migrants, three-quarters were involved in farm work before their move, but almost all had primary education, with 42 per cent going on to complete secondary level too. As almost one-quarter of the population in the

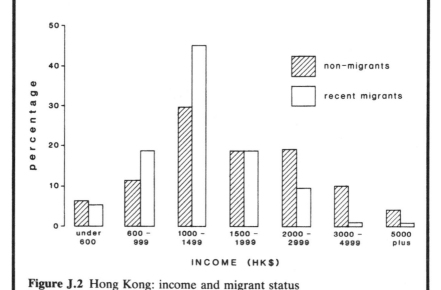

Figure J.2 Hong Kong: income and migrant status

Case study J *(continued)*

neighbouring province are illiterate, the migrants to Hong Kong clearly follow the usual Third World pattern of being positively selected with respect to education.

Plate J.1 Hong Kong peripheral squatter settlements are constantly refuelled by a steady stream of Chinese migrants. Their permanence has been confirmed by recent government moves to provide basic facilities.

Case study J *(continued)*

Again, as has been found elsewhere, although level of education has a positive correlation with propensity to migrate, it does not seem to influence employment to the same extent. Although few found difficulty in obtaining wage employment, jobs were usually at the lower end of the scale. Almost three-quarters were in some sort of blue-collar work and the migrant income profile is distinctly different from that of the resident population (Figure J.2).

As migrants do not qualify for public housing, at least not without a lengthy wait, many tend to concentrate in the older tenement districts of the metropolitan area where relatively cheap if small units of accommodation exist; or else they establish themselves in the squatter settlements that still encircle Kowloon (Plate J.1). Indeed, it is alleged that it is immigration which keeps the squatter population so high in Hong Kong (still in excess of 250,000 despite 30 years of intensive public housing programmes).

Although the nature of the migrant population, their background and impact in Hong Kong have clear parallels with rural-to-urban migrant streams in other parts of Pacific Asia, the flow of people from the neighbouring regions of South China into the colony has been much more strongly subject to political influences than elsewhere. Although a prosperous city will always prove to be a magnet for migrants from an impoverished agricultural hinterland, the political separation of Hong Kong from China, in terms of economic philosophies, has made the contrast much deeper. Furthermore, the political climate between Hong Kong and China has enormous impact on the dimensions of the flow of both illegal and legal migration. For example, the improvement of relations and relaxation of controls following the demise of the Cultural Revolution was an important factor in the upsurge of migrants in the late 1970s. In contrast, the cooling of relations following the anti-democracy purges of 1989, has caused an increase in emigration from Hong Kong itself as uncertainties over 1997 re-emerge.

In short, although migration to Hong Kong appears to have

Case study J *(continued)*

> much in common with rural-to-urban migration elsewhere in
> Pacific Asia, the international political dimension is the ultimate
> arbiter of events in the territory.
>
> *Source*: Sections of this case study are based on research by Ron
> Skeldon (see Further reading).

Indonesian population movements

Geographical and historical background

The sheer size of Indonesia tends to be overshadowed by the Asian
giants of China and India, but Indonesia is big by any standards. Its
population of 170 million makes it the fifth largest country in the world,
and the largest Islamic state; whilst geographically its 13,000 islands
extend over almost 2 million square kilometres. In comparison to many
other developing countries, Indonesia appears to be relatively rich in
resources – land, mineral and human. Unfortunately these are very
unevenly distributed with Inner Indonesia (Java, Madura and Bali)
having most of the population and the physical resources being scattered
over the outer islands. In addition, Indonesia is deficient in management
and organizational skills, infrastructure and finance and is one of the
poorest countries in Pacific Asia.

Many of Indonesia's problems undoubtedly stem from its colonial
past. The Dutch ruled Java for some 300 years and the other islands for
almost a century. It has been argued that the Dutch were the least
complex of the colonial powers – not seeking empires, Christian
conversions, materials for industry or markets for their own goods,
simply material for trade. This did not make the Dutch any less
rapacious, however, and colonial Indonesia was dominated by plantation agriculture, particularly sugar, rubber and tobacco. Indonesians
were seen primarily as plantation labour so that most local commerce
was in the hands of immigrant Chinese.

After a somewhat protracted and bloody independence struggle,
Indonesia became a republic in 1950 and for the next 15 years was led by
Sukarno, one of the liberation leaders. His government was socialist (in
rhetoric at least) and he nationalized most Dutch enterprises. However,

the state did not have the resources to sustain production or guarantee cheap labour (as in the colonial period). Small-holders also suffered through lack of government support and the quality and quantity of exports fell. The ensuing economic crisis precipitated political upheaval and Sukarno was replaced in a complex, bloody, communist/military coup by the present government led by Suharto.

The new government was pro-western and favoured modernization by promoting mineral exports and industrialization. Investment funds flowed into Indonesia, especially from Japan (see Chapter 3), which is now the country's leading trading partner (45 per cent of all trade). Until the 1980s the economy was expanding rapidly at around 8 per cent per annum but global recession, particularly the drop in oil prices, has cut the rate of growth by more than half. In an effort to improve this situation, the government is looking towards downstream processing of its own minerals to increase value added profits.

What has tended to take a back seat in recent development priorities is the rural sector, despite the fact that more than three-quarters of the population still live in rural areas and agriculture contributes over half of Indonesia's GDP. In the most crowded areas of western and central Java, rural population densities average over 500 per square kilometre and in parts they reach 2,000. The practice of subdivision of inherited landholdings, together with the increased penetration of commercial agriculture, has meant that many rural households subsist on tiny plots of land or else sell their holdings and try to obtain waged employment (on commercial plantations or in rural non-farm work). Recent data have revealed that over one-third of all rural households now have no land of their own and for more than half, their holdings amount to less than 1 hectare

Urban migration

It is hardly surprising in these circumstances that many choose to move, particularly from the most crowded areas, and try for work in the cities. Urban population growth rates in Indonesia are amongst the highest in Pacific Asia and are increasing. In the 1960s the average annual urban growth rate was 3.9 per cent; by the late 1970s this had risen to 4.5 per cent and it currently stands at 5 per cent. Although much of the urban growth in absolute terms is concentrated in the largest cities in Indonesia (one-third occurred in 'millionaire' cities), this is to be expected given their numerical dominance. What is much more important is the fact that the high rate of increase is the same

throughout the entire urban hierarchy. Indeed, the rate of growth is
slightly higher for cities of between 100,000 and 500,000. Part of the
explanation for this across-the-board increase lies in the fact that
Indonesia is a fragmented state so that each island will have its own
capital, in effect its own primate city. Indeed, if we examine the figures
for the larger centres on islands such as Kalimantan, the rates of urban
growth are over half as much again as on Java or Sumatra.

All this is even more remarkable when it is remembered that many
intermediate towns and cities are themselves sources as well as destinations for migration. This means that the extent of population turnover in
such settlements is very high indeed. In fact, the migration process in
Indonesia has become more complex in general over the last decade or
so, largely owing to the improvement in communications, most notably
in regional bus services. This has markedly increased the propensity to
migrate on a temporary basis, either in conventional commuting, or in
what has become known as circular migration. In the latter, the village
home is retained but the migrant stays in the city for several months at a
stretch, attempting to earn as much money as possible. The impact of
these structural changes in migration patterns has been considerable as
Graeme Hugo's work in Indonesia reveals. Some of his data for West
Java indicate how the region around Jakarta has been affected by this
change in mobility patterns (Figure 5.2).

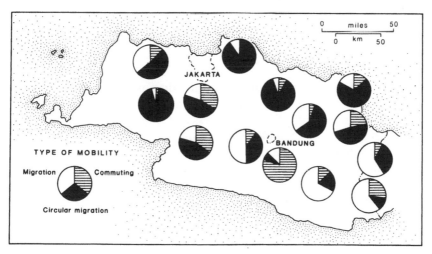

Figure 5.2 West Java: mobility patterns to Jakarta

The implication of such trends is that permanent migration is coming from an ever-widening area and, indeed, this is the case. Figure 5.3 reveals that many migrants to Medan in North Sumatra come from Central Java, well over 1,000 km away. By and large longer-distance moves are made by the more educated migrants who feel that they have something to offer, although it is also true that migrants in general tend to be better educated, but have less access to land and other family resources.

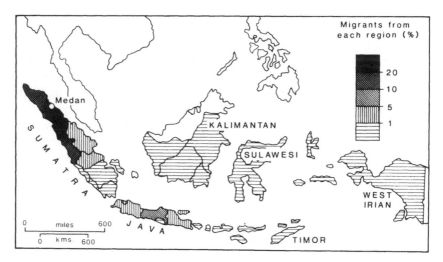

Figure 5.3 Indonesia: migration to Medan
Source: Ulack and Leinbach, 1985

As yet, Indonesia does not display any dominance of females within the migrant stream, even in the 15–25 year age bracket; a reflection of the fact that its industrialization process is still far behind that of Taiwan or South Korea, for example. Thus, in the migration process, males still outnumber females by about two to one, and move to the city when they are young and unmarried in search of work.

The type of work which a migrant is likely to obtain in the city varies enormously according to the economic structure of the city concerned. In the big cities of Pacific Asia's newly industrializing countries, the majority of the migrants may expect to find work in factories – if not immediately then within a short time (see Chapter 9). In other cities, the informal or petty commodity sector still plays an important role in absorbing migrant labour. For example, in the case of Medan cited previously, as many as 60 per cent of the migrants find jobs in the informal sector.

Interestingly, the educational background of the migrant does not appear to influence employment on arrival in the city. More education may increase the propensity to migrate but it does not guarantee a better job in the city. Not that informal sector work is necessarily poorly paid; the difference between the average monthly wage in the two sectors in Medan was only about 20 per cent.

For most migrants the move to the city is usually set within a network of relatives and friends who not only give advice on planning the move, but act as contacts for employment and shelter immediately following the move itself. In fact, in Indonesia in particular, this relationship between the rural and urban components of the migrant stream can be very narrow and specialized. Table 5.1 shows the results of a survey by Graeme Hugo in West Java and indicates clearly the degree of job specialization of migrants from various villages.

Table 5.1 West Java: occupational specialization in urban areas of village migrants

| | | Proportion of all working migrants | | | |
Village	Main occupation	%	Second occupation	%	Total %
1	Ground-nut hawker	65	Government/army	15	80
2	Cooked-food/cigarette hawker	35	Day labourer	22	57
3	Cooked-food hawker	43	Jewellery hawker	21	64
4	Pedicab driver	57	Day labourer	16	73
5	Pedicab driver	41	Factory worker	34	75
6	Labourer	35	Hospital worker	13	48
7	Kerosene hawker	32	Household domestic	15	47
8	Airline/hotel workers	32	Household domestic	10	42
9	Kitchen utensils hawker	60	Government/army	12	72
10	Driver	27	Government/army	26	53
11	Pedicab driver	38	Construction worker	20	58
12	Carpenter	49	Government/army	28	77
13	Barber	31	Bamboo worker	20	51
14	Bread hawker	42	Driver	32	74

Often this is the result of a single person obtaining employment in a specific field, such as the airport. As that person becomes well established and perhaps obtains a more responsible position, he is in a position to help others from his family and village to obtain work. On many occasions too, this occupational specialization in the city is tied to a specific place of residence. This is particularly true of circular migrants who do not want permanent residence in the city and tend to live in lodging-houses (*pondoks*), again with people from the same village and usually related to a single occupation.

The response of the Indonesian state

Most governments do very little about rural to urban migration. They profess to be concerned about the build-up of an urban underclass but know full well that its existence is essential for the cheap labour which constitutes one of the main attractions to foreign manufacturers. In this sense the informal or petty commodity sector is seen as performing a useful function in absorbing migrants and giving them some access to employment and income but, at the same time, constituting what Marxists would call a reserve army of labour, ready to be recruited into formal capitalism when the need arises.

Nevertheless, the build-up of large numbers of urban poor is seen by some, particularly by international agencies, as a threat to the political stability which is essential to capitalism. Agencies such as the World Bank are therefore keen to recommend and support programmes dealing directly with the growth of urban poverty. In part, such programmes involve the provision of basic needs within the city, but these are expensive and appear to be social overheads bringing little return in comparison to investment in economic growth. Thus many policies which 'deal' with urban poverty have tended to adopt an alternative approach, which is to try to prevent it building up beyond 'acceptable' proportions.

A positive version of this alternative approach would be to make rural areas and rural life more attractive, economically and socially, so that fewer people migrate. Such regional and rural development programmes have already been discussed in Chapter 4 and it is clear that they are, in general, poorly structured and ineffective in restraining migrations – quite the reverse in some areas. Thus, policies to curb rural to urban migration tend to be negative rather than positive. These range from direct constraints on migrants entering cities to measures intended to deflect migrants away from urban areas either overseas or towards other rural areas within their own country, if suitable areas with development potential exist.

Indonesia has undertaken measures both to curb and direct rural to urban migration. In fact, Indonesia's rural resettlement programme is the longest established of such schemes in Pacific Asia. Transmigration, as it is known, began during Dutch colonialism, and in theory was intended to reduce population densities in West and Central Java by transferring peasants to Sumatra. It seemed more than coincidental that labour-intensive plantations were being opened up in Sumatra at the same time that the scheme came into operation. In the four decades that the Dutch operated the scheme, only about 200,000 migrants were moved from Java and this had a negligible effect on population densities.

Since independence some 5 million migrants have participated in the transmigration programme, almost all of whom were moved during the two five-year plans since 1979. Although the principal focus for migration continues to be Lampung Province in southern Sumatra, destinations have broadened substantially during the 1980s to include border provinces, such as Kalimantan and West Irian, where it has been deemed necessary to augment the Javanese population.

Although those selected to participate in the transmigration programme are usually very poor households from overcrowded areas of high environmental hazard, the impact on the areas of origin and reception, and on the migrants themselves, have been disappointing. West and Central Java remain very poor, with 47 per cent of their populations below the poverty line, almost double the national average. The migrants receive tiny farms in marginal areas and overall although they constitute 3 per cent of total population, they produce only 0.4 per cent of the country's rice. Indeed, there is considerable evidence that their attempts to expand production in these marginal areas have resulted in accelerated environmental deterioration, usually following removal of forest land.

Overall the transmigration programme seems to have taken on more of a political than a demographic or economic rationale, being seen at present more as a strategic way of securing difficult frontiers in lands not dominated by ethnic Malays. Thus the West Irian province is scheduled to receive one million Javanese transmigrants to add to a Melanesian population of 1.2 million. This appears to be having the reverse effect to the one intended, since the Melanesians are increasingly agitating for separatism or union with neighbouring Papua New Guinea.

Following the recession of the mid-1980s, transmigration had virtually stopped by 1987. With the principal financial backer, the World Bank, becoming increasingly concerned about the environmental problems, there have been signs that the programme has shifted into a 'consolidation' phase. Whether the resettlement momentum will ever restart is difficult to say at present.

The demise of the transmigration programme will undoubtedly put more pressure on the large cities of Java as potential migrants shift their target. This may induce a response from the authorities similar to that of the early 1970s, when Jakarta was declared to be a closed city and a pass was deemed necessary for official residence. Such a pass, or *kartu*, was not issued unless permission from the home administrative unit was given, a job and residence in Jakarta could be guaranteed, and the return fare could be guaranteed. Even so the system failed. *Pondoks* gave the necessary security of residence and employment, and even if other conditions were not satisfied, a *kartu* could always be obtained on the black market or a bribe paid so that officials or police ignored its absence. It was worth paying whatever was necessary to gain access to the higher wages and greater job opportunities of the capital.

The failure of both the pass system and transmigration to curb rural to urban migration means that Jakarta and other large cities are growing faster than ever in a country where the urbanization process has only just begun to accelerate. Rural to urban migration and the corresponding problems this presents for urban and national authorities is likely to worsen considerably for the foreseeable future in Indonesia – what political repercussions this will have remains to be seen.

Key ideas

1 Although population growth rates are lower in Pacific Asia than in the Third World as a whole, there is considerable variation both between and within countries.
2 Internal and external labour movements continue to be important in Pacific Asia.
3 Rural-to-urban migration remains the largest type of population movement.
4 In many countries, state governments have sought to control both population growth and population movement.

6
Ethnic plurality and development in Malaysia

Ethnic variations and population mobility in Pacific Asia

The degree of ethnic variation within Pacific Asia varies enormously but in general the situation becomes more complex from northeast to southwest. In a way this is not surprising since it is in the Southeast Asian region that most population movement has occurred. In the past such movement affected the mainland and the islands in different ways. On the former, Vietnamese, Thai, Khmer and Burmans contended for dominance as economic or political situations changed. Within the Southeast Asian archipelago the dominant Malays were supplemented and strongly influenced by relatively small numbers of Arab and Indian traders. Indeed, the major influences in pre-colonial Southeast Asia were from cultures that did not necessarily have a large physical presence, viz. the Indian and Chinese governmental structures and the Islamic religion.

However, during the long mercantile colonial period, as trade and commercial activity increased, there emerged a growing Chinese presence in the cities of the region. Whilst indigenous populations tended to concentrate on commodity production, the local assembly and transport of these commodities were increasingly assumed by expatriate Chinese who had both the inclination for, and experience of, this work. However, few regarded themselves as permanent residents of the ports in which they operated; their loyalties lay firmly with their ancestral homeland.

Clearly the efficiency of the Chinese business community impressed the European powers and with the economic intensification of the industrial colonial period, the commercial role and demographic presence of the Chinese community expanded enormously. Even colonial newcomers, such as the French, had no hesitation in inviting Chinese into Indo-China to manage commercial activities such as rice milling and retailing.

However, the colonial period witnessed much more substantial mixing of populations within the Southeast Asia region. Primarily, this centred around the new labour needs of both plantation and mining enterprises. In many instances, indigenous populations were either unsuitable or too few in number for such work, even when 'encouraged' to enter the cash economy through tax demands. The response was to contract labour in large numbers from other areas. Thus, northern Vietnamese migrated to the south, Tamils were brought over from India, whilst unskilled labour movements out of China accelerated enormously (see Case study K).

For the most part, such large-scale movements were more characteristic of the period up to the 1920s and began to fade with the onset of recession. They were then supplemented by other forms of population movement designed to even out development within the colonies (see Chapter 5), a policy which has continued through to the present day in schemes intended to open up new lands for settlers from more crowded areas.

Of course, as noted in earlier chapters, such movements have been overtaken by urbanization as an agent in population redistribution. Although this might seem to be less relevant to the question of ethnic mixing than the international movements discussed above, it has proved to be very significant in countries where urban and rural populations had a separate ethnic character.

All of the above population and ethnic changes have occurred in Malaysia and the legacy of several hundred years of population shifts has been a complex cultural and demographic situation that, more than anywhere else in Pacific Asia, has both shaped and been influenced by the development process. Race, class, politics, economic change and social pressures are indivisible and omnipresent in Malaysia.

Case study K

Burma: An ethnic kaleidoscope

Of all the nations in the region covered in this book, Burma is ethnically the most complex. Even more so than Malaysia, its future is threatened by confrontations that not only have a fundamental ethnic basis but are also overlain with colonial, cultural and political divisions, none of which correlates. To appreciate fully the present situation, it is necessary to begin in the pre-colonial period.

The major ethnic group in the country is the Burmans who comprise some two-thirds of the population and occupy the main Irrawaddy plain (Figure K.1). They had settled in this area by the ninth century AD and over the next 400 years constructed their great capital at Pagan, the pagodas of which constitute one of the world's principal architectural monuments. As with other pre-colonial civilizations, the political economy was based on the rich, rice-producing lands of the delta.

In the mountains and valleys around the Burman core are the multitude of minority groups, most of whom overflow into the surrounding countries of Bangladesh, India, China, Laos and Thailand. Of these, the largest and most autonomous group is the Shen; Thai-speaking and Buddhist, they occupy about 20 per cent of the land area of Burma but comprise only about 10 per cent of its population. The other leading minority groups (the Karen, Chin and Kachin) are socially, politically and culturally more fragmented, despite their general geographical clustering. Many are animist, although there are also substantial Christian and Buddhist populations. Muslim minorities can be found along the west coast, close to Bangladesh.

Relationships between these groups were flexible in the pre-colonial period. The Burmans exercised a loose political control but never sought to impose military domination. There was considerable trade between the various groups, but little cultural interaction. Although occasional warfare erupted, there is little evidence of wholesale killing. As in much of Pacific Asia at this time, land was plentiful; it was people that were scarce.

Isolated by the surrounding ring of mountains, Burma's fluctuating

Case study K *(continued)*

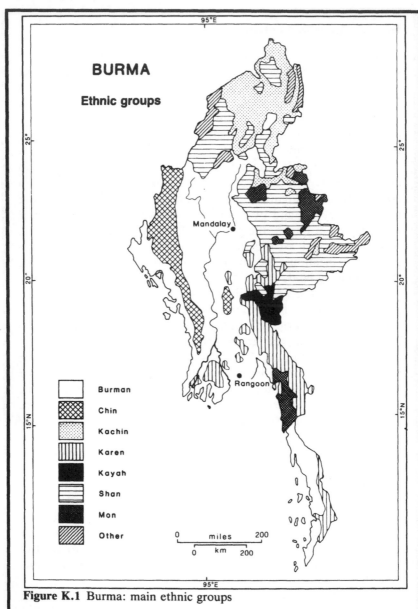

Figure K.1 Burma: main ethnic groups

Case study K *(continued)*

politics have mostly looked inwards rather than outwards and it may well have been the case that eventually the Burmans would have established a mature and stable relationship with the various minorities. Unfortunately, this process was interrupted by British colonization in the nineteenth century which effectively froze internal ethnic relations as they were until the 1940s. Indeed, by politically favouring the predominantly Christian Karens, the British sowed the seeds for post-colonial antagonism.

After annexation in 1886, the British also introduced further ethnic complications, as in Malaya, by encouraging the immigration of other Asian peoples. The economic growth of the colonial period was firmly built on the back of immigrant Indians. Agriculture expanded through Indian labour and tenant-farming; the credit for expansion came from Chettyars (Indian moneylenders); and the middle class in general became dominated by Indians and, to a lesser extent, by Chinese. By 1931, although Indians numbered only 7 per cent of the total population, they occupied over one-third of mining jobs, about 15 per cent of the jobs in trade and industry, 27 per cent of those in the civil service and 43 per cent of transport and police employment. In particular, Rangoon became an Indian city and by 1901 Indians comprised 53 per cent of the population compared to only 32 per cent of Burmese. Colonial and Hindu architecture dominated the public areas of the city. In short, there was a clear ethnic division of labour, with the British at the top, the Indians and Chinese in the middle and the Burmese very much at the bottom.

The world economic recession of the 1930s caused the price of many of Burma's exports, particularly rice, to drop dramatically. The economic hardship this created fuelled political resentment of foreign domination, and banditry became rife in the country. In addition, as in many other colonies during the late colonial period, the seeds of socialism were germinating amongst those Burmese who had been educated in the political hothouse of Europe.

The Second World War itself added to the already complex ethnic, cultural and political mess that comprised colonial Burma because the Burmans sided with the Japanese, whilst the Karens fought with the British. Although the Burmese reunited to help

Case study K *(continued)*

oust the Japanese in 1945, another dimension had been added to the divisions within the country. The Karens expected independence in return for the loyalty they had shown the British but, in 1947, all of the states were formally incorporated into the independent, Socialist Union of Burma.

The term 'union' was as far from reality as it was possible to be. The old, loosely structured but effective ties of the pre-colonial period had vanished with the monarchy. Instead, anti-colonialist and anti-Japanese struggles had armed the minorities and encouraged thoughts of independence. However, not only were there regional ethnic uprisings against the new Burman-dominated state in the late 1940s, there was also political dissent. Two separate communist groups were in active opposition, whilst the ruling socialists were also split. To complicate matters further, a Muslim separatist movement began along the west coast, whilst fleeing units of Chiang Kai-shek's nationalist Chinese also entered the country in the northeast, where they reputedly became actively involved in the military training of minority groups.

Gradually, the central government acquired control and developed a state based politically on a fusion of socialism and Buddhism, and ethnically and economically on the Burman lowlands. However, when in 1961 Buddhism was made the official state religion, many of the minority groups began to agitate once again for greater autonomy and even secession. This induced a military coup and an intensification of the Burmese way to socialism, assuaging indigenous groups by nationalizing economic enterprises, taxing savings and forcing many Indians to leave the country. Meanwhile anti-Chinese riots in Rangoon also resulted in more active Chinese support for communist groups in the border areas.

Over the years, the socialism of the ruling faction has been modified into a more pragmatic version, in which agriculture rather than industry forms the mainstay of the economy, and in which Burmese international isolationism has diminished. New states have been created giving seven internal minority divisions and seven divisions within the Burman areas. However, this balance is illusionary, since political power, economic growth and social progress remain centred on Rangoon (Figure K.2).

Case study K *(continued)*

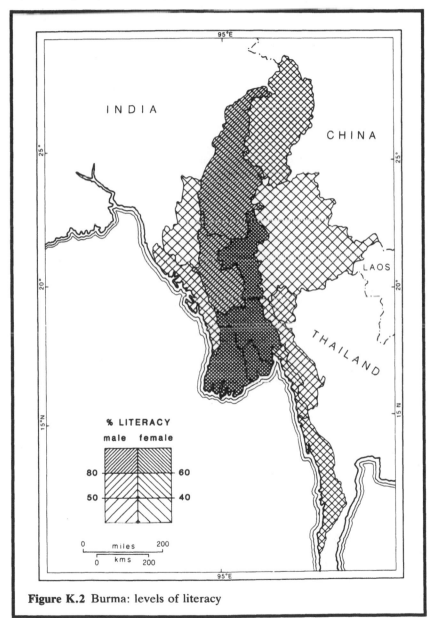

Figure K.2 Burma: levels of literacy

Case study K *(continued)*

The result of this long and complex history is that the Union of Burma exists only in name. An intricate and mobile kaleidoscope of ethnic, cultural, religious, political and economic fragments exists within the peripheral regions of the country. Moreover, the wealth earned from mineral, hardwood and opium production has ensured that this separatism is sustainable at least for the foreseeable future, forcing the Burmese government to spend one-third of its annual budget on internal defence. Until the government begins to look outward in its development policies, this situation is likely to persist.

Sources: David Steinberg (1982); Chris Dixon (1991); George Orwell (1934) *Burmese Days*, Harmondsworth: Penguin.

From Malacca to Malaysia: the evolution of the modern state

Before the Europeans arrived, the Malay peninsula comprised a series of small riverine sultanates of which Malacca was by far the most important, controlling western access to the spice islands. As its importance grew, despite a poor natural harbour, so it became an important focus for east–west trade within Asia – the point at which Chinese and Indian goods were exchanged. During all this time it developed a heterogeneous community of Arabs, Malays, Chinese, Javanese and other ethnic groups, so that when the Portuguese assumed control in 1511, Malacca mirrored the racial complexity that the whole peninsula would eventually follow.

For the rest of the sixteenth century, Malacca continued to dominate both intra-regional and inter-regional trade, being an important link in the chain of Portuguese ports from Goa to Macao. But by 1600 the rival Dutch port at Batavia (now Jakarta) had taken away much of its political and commercial pre-eminence. In 1641 the Dutch had assumed control of Malacca (Plate 6.1) but it never regained its importance and was well past its best by the time the British took over in 1824.

By this time the tin deposits on the peninsula were known but it was not until the mid-nineteenth century that these were more fully exploited. By this time Britain had acquired Penang in 1786 and Singapore in 1819, two island bases which were used far more

Plate 6.1 Dutch Malacca: a century and a half of Dutch dominance has left few architectural signs

extensively than Malacca to tap the resources of Malaya, particularly after all three ports were drawn together in the Straits Settlements in 1867. This formalized a wide range of *ad-hoc* agreements and trading arrangements, particularly relating to tin which by now was in great demand in industrializing Europe.

Tin mining was largely financed by Chinese commercial capital which imported its own labour directly from China. The rich profits to be made from tin soon led to quarrels between Malay landowners and Chinese commercial interests, threatening to disrupt the supply. In 1874 the British established protectorates over the main tin-producing states, installing residents to advise (and rule). Gradually control was extended to all of the states and by 1914 the Federation was complete. Similar events in North Borneo had also resulted in British 'protection' being extended to what is now East Malaysia.

The new Pax Britannica stabilized development and accelerated both tin mining and the in-migration of Chinese; between 1880 and 1910

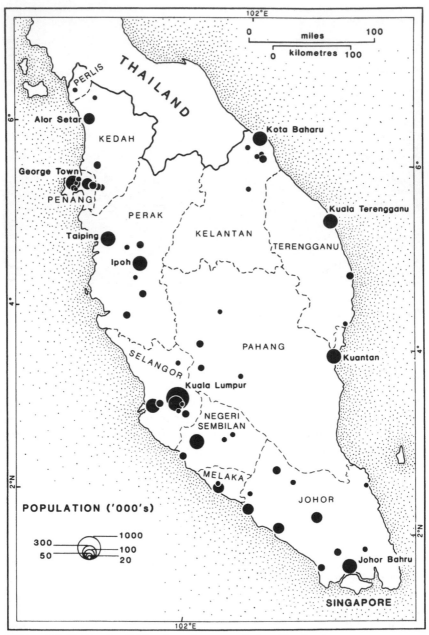

Figure 6.1 Peninsular Malaysia: urban population distribution

some 6 million entered Malaya. Until the Second World War the great majority of the Chinese did not consider themselves to be permanent settlers, although that is what many eventually became. This goes a long way to explaining the lack of any specific policy towards the growing ethnic association with certain activities and areas. Many of the industrial towns that emerged in the tin-mining areas in the western part of the peninsula were overwhelmingly Chinese, including the new central administrative centre of Kuala Lumpur (Figure 6.1).

The road and rail infrastructure that facilitated tin mining and export also opened up the interior of Malaya for commercial agricultural development. But it was only in the 1890s with the introduction of rubber trees, in response to massive acceleration in world demand, that agricultural exports became important. By the First World War some 2.5 million acres of tropical rainforest had been replaced by rubber plantations.

The British colonial government made a conscious decision to keep Malay peasants in food production, primarily paddy rice, and in 1913, after others had taken possession of the more commercially valuable land, Malay peasant rights were confirmed in most of the remainder. This left the British companies, which dominated the rubber industry, in search of a workforce. They solved their problem by recruiting indentured Tamil labour from South India where many of the British managers had previously served. At the peak of this migration immediately prior to the First World War, some 100,000 Tamils were entering into Malaya each year.

Economic development had thus been tied very closely to ethnicity, with the Chinese involved in urban/mining activities, the Indians providing plantation labour and the Malays, both rich and poor, owning or working the remaining agricultural land (Table 6.1). Ethnic differences and rivalries inhibited unification along class lines and the British were left with a very profitable, easy to administer colony. Although all urban centres had grown, the primate city and effective capital was Singapore, through which two-thirds of Malaya's trade passed. Like other towns, however, it had always been a predominantly Chinese city. Over 61 per cent of its 50,000 population in the mid-1840s were Chinese and this had risen to 80 per cent by the 1930s.

During the inter-war period, international prices for tin and rubber fell and many estates diversified into oil-palm, pineapples and coconuts. But the end of the primary export boom also began to spread discontent amongst the Malays because not only were they the poorest of the three

Table 6.1 Malaysia: employment and ethnicity

	% Labour composition 1930			% Labour composition (1970) 1980		
	Estates	Mines	Padi	Primary	Secondary	Tertiary
					*	
Malays	3	1	97	(68) 66	39 (38)	47
Chinese	23	92	1	(19) 20	51 (51)	42
Indians	74	7	2	(12) 13	9 (10)	11

Source: Mehmet (1986)
Note: * Secondary and tertiary employment combined

racial groups but had almost become a minority in their own country. Large-scale labour immigration was discontinued but by now many of the immigrants had changed allegiances and regarded Malaya as their home. The Chinese in particular began to demand better working conditions and equal political rights. It is, however, crucial to appreciate that many of these objectives were class specific – thus the wealthy Chinese were more concerned with political rights, and the Chinese proletariat with better working conditions, so that socialist union movements gained considerable ground between the wars.

The 1930s also saw a similar politicization of the Malays in response to these pressures from immigrant groups. The educated elite wanted special rights in terms of access to administrative positions. Most of the Malay peasantry, however, remained politically docile. Thus what appeared to be an ethnic conflict between immigrant and indigenous groups over political and economic privilege (not power which was still in British hands) also had a clear class dimension.

The Japanese occupation may have destroyed the myth of European superiority but it stirred the ethnic pot more vigorously. The Malays were elected to (nominal) political leadership whilst the Chinese had their businesses confiscated or destroyed. The growing opposition to Japanese colonialism thus became predominantly Chinese and politically leaned heavily to the socialist/communist left. When the British returned, this insurgency intensified, particularly as the Malays were once again favoured for administrative posts. Two of the largest employers of Malay (and to a certain extent Indian) labour were the police and army who undertook the brunt of the military actions against the communists. What was intended to be an anti-colonial struggle, therefore took on an ethnic character. The principal outcome of the insurgency of the 1950s was the destruction and repression of the trade union movement (because of its association with the Malayan Communist Party), and the entrenchment of Malays in public administration, the army and the police.

The communist insurgents were thus hindered not only because they could not claim to be a nationalist movement, compared to other countries in Pacific Asia, but also because they comprised only a class fraction of the Chinese community since wealthy, middle-class Chinese wanted little to do with revolution. Moreover, Britain had already determined on independence, which was granted to peninsular Malaya in 1957. Singapore then joined and left, but the addition of the two north Bornean states of Sarawak and Sabah had effectively created modern Malaysia by the mid-1960s.

Years of illusion: 1957 to 1970

The British withdrawal from Malaysia had the effect of removing the lid from a simmering pot. Economic mismanagement during the 1960s effectively turned up the heat, so that by the end of that decade the situation was ready to boil over.

Essentially the British left the Malay elites in political control but without economic power, whilst the wealthy Chinese had economic wealth but little political power. Below these elitist groups were a Chinese petty bourgeoisie and proletariat, and a massively poor Malay peasantry. The Indian community still dominated rural labour and had also penetrated the petty bureaucracy and rural commerce.

The new government, advised by the World Bank, continued to base its economic growth on primary commodity exports with some diversification into import-substitution industry. This was orthodox capitalist development strategy for the Third World at the time. Again under World Bank advice, foreign capital was to be encouraged to invest in the country in order to achieve these objectives. In order to make such investment attractive, the government invested heavily in infrastructural projects (improved transport, land clearance schemes), subsidies, or tariff protection schemes.

The main problem with the 1960s was not that economic growth failed to materialize but that the distribution of the benefits was poorly managed. For example, in the industrial sector, although a large number of branch plants were established, so that by the 1970s over 90 per cent of consumer goods were being produced 'locally', this did not create much in the way of new employment as the domestic market provided by a population of less than 10 million was comparatively small. The benefits of industrial growth thus accrued to foreign firms, and many of those who migrated to the cities in search of work (mostly Malays) were disappointed.

In the agricultural sector, the government invested substantial amounts of money in infrastructural and technological improvements for the plantation sector of the economy. As a result, production per worker more than doubled between 1960 and 1970. This was an important factor at a time when the world prices were falling, so that the estate sector still contributed 40 per cent of Malaysian export earnings in 1970. However, improvements in productivity were not reflected in wages which in real terms remained relatively static (Figure 6.2).

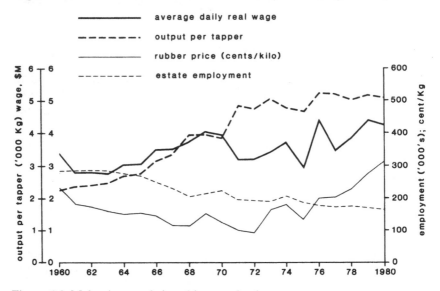

Figure 6.2 Malaysia: trends in rubber production
Source: Mehmet 1986

Not all of the major agricultural exports were produced by the estate sector; 40 per cent of rubber, for example, came from small-scale producers of less than 100 acres. Overall production from small-holders slowly expanded during the 1960s, as did rice production, as clearance schemes made more land available. However, this overall expansion concealed the tiny size of most holdings which averaged 2.4 hectares, ranging from 3.5 hectares for Chinese to 2.0 hectares for Malays. Moreover, the subsidies offered to both rice and rubber producers encouraged amalgamation of small-holdings, so that by the end of the 1960s the average size of owner-occupied units had risen whilst that of tenants had fallen. In short, most

Table 6.2 Malaysia: inequality and ethnicity

	Average Income (M$)			Gini Coefficient*		
	1957	1970	1980	1957	1970	1980
Malays	140	171	513	0.34	0.40	0.47
Chinese	302	381	1094	0.38	0.46	0.58
Indians	243	301	776	0.37	0.47	0.49
All	215	246	763	0.41	0.50	0.57

Source: Mehmet (1986)
Note: * The Gini Coefficient is a measurement of income distribution. It ranges between zero (total equality) and one (maximum inequality).

small farmers, who were primarily Malay, were poor and getting poorer.

The end-product of the 1960s was a steady growth and diversification in national economic development but growing inequality in the distribution of that wealth (Table 6.2), largely due to structural reasons. The poor were poor because they received low wages or were deprived of land. The consequence was a growth of, largely Malay, migration into the cities where insufficient jobs had been created and unions were still banned. Inequalities thus increased in many directions – between rural and urban areas, between races, between classes within racial groups. As a result, interclass and inter-ethnic tensions escalated. Malay peasants blamed their worsening situation on those in the cities (i.e. on the Chinese); the poor urban Malays resented Chinese domination of formal and informal wage employment and commercial activities; the Chinese proletariat in their turn blamed government repression of trade unions for their low wages; and the Chinese bourgeoisie, both petty and *haute*, chafed at their continued exclusion from government and at favouring of foreign manufacturing firms.

The pot finally boiled over following the 1969 elections in which Chinese opposition parties won control of the two main urban states of Penang and Selangor. This appeared to threaten the political and economic future of both rich and poor Malays. Some Malays in Kuala Lumpur and several other cities 'ran amok', Chinese gangs retaliated and several hundred people were killed within a few weeks of communal violence.

Years of disillusion: the New Economic Policy since 1970

The Malaysian government placed the blame for the inter-communal violence squarely on economic inequality between the races and formulated its New Economic Policy (NEP) in response. The first

Plate 6.2 Kuala Lumpur: in the 1970s old and new jostled for attention in one of the most rapidly expanding capital cities in Asia

objective was to eradicate poverty amongst all races, but as most of the poorest households were rural Malays, this implied a built-in bias. Indeed, this was firmly articulated in the second principal goal, which was to establish greater equity in control of economic growth; in particular it was intended to increase the Malay share of ownership and employment in all sectors of the economy to 30 per cent. The shift was ostensibly to be at the expense of foreign enterprises.

This amounts to an affirmative action programme in favour of the Malays. Ozay Mehmet claims that historically the Malay masses have always given their rulers total loyalty and in return they have expected protection. During the 1960s it appeared to Malays that this relationship was being eroded. The NEP restored the *status quo*.

The major economic strategy was, therefore, to abandon *laissez-faire* capitalism in favour of an increased role for the state in the allocation of resources, in the regulation of the economy and in direct investment in both public and private enterprise. The tactics revolved around the favourable treatment of the Malays through scholarships, subsidies, quotas, licensing and trade concessions. All this has to be set against an international scene in which MNCs were actively searching for new investment opportunities to promote industrial growth. The combination of a strong, receptive central government and an increased volume of overseas investment has meant that there has been a sizeable spatial and sectoral shift in the economy away from primary commodity production towards urban manufacturing (Table 6.3). It has also meant that the foreign sector of the economy has been difficult to erode, so that Malay advancement has largely been at the expense of non-Malay domestic investment interests, i.e. the Chinese.

Table 6.3 Malaysia: selected indicators of development

	1960		1987	
Population	7m	(2.4)	16.5m	
GNPpc US$	1100	(4.3)	1810	(4.1)
Urban %	25	(3.3)	40	(5.0)
Urban % in KL	19		27	
Labour force % in				
Agriculture	63		42	
Industry	12		19	
Services	25		39	
GDP %				
Agriculture	37		20	(3.6) [3.4]
Industry	18		37	(5.5) [−2.0]
(Manufacturing)	(9)		(20)	(4.9) [−3.0]
Services	45		43	(7.9) [7.0]
Merch. imports %				
Food	29		10	
Fuels	16		6	
Other primary	13		4	
Machinery, transport equip.	14		50	
Other manufactures	28		30	
Merch. Exports %				
Fuels, minerals, metals	20		25	
Other primary	74		36	
Textiles, clothing	–		3	
Machinery, transport equip.	–		27	
Other manufactures	6		13	

Notes: ( ) average % rate of growth per annum 1980–88; [ ] % growth in 1988
Sources: WB Development Report (1989), Cho (1990)

Overall there is no doubt that the Malaysian economy has experienced steady and impressive growth since 1970 (Table 6.3). But Malaysia is a resource-rich country with a small population, so that such growth should not be surprising. The question is whether the benefits of growth have been fairly distributed, even by the Malaysian government's own interpretation of events. Although the number of rural dwellers living in poverty appears to have been substantially reduced, such generalized trends mask important considerations. The first of these relates to the fact that, for most rural dwellers, poverty remains as bad as it was in 1970. In the rubber plantation sector, for example, 58 per cent of estate workers were estimated to be in poverty in the mid-1980s, a fall of only 1 per cent over 1970. The reason is that commercial plantations sought to counteract falling world prices by reducing output but increasing productivity per worker (Figure 6.2). This was achieved not only by increased infrastructural investment by the state but by holding wages steady in real terms. As a result the wage differential of tappers and manufacturing workers widened from 1.9 to 2.9 between 1967 and 1981.

In the smallholder sector, whether in rubber or rice production, the situation is not much better. Holdings remain uneconomically small, particularly for tenant farmers, and the benefits of increased government investment in land clearance, subsidies, and new technologies have been felt by relatively few larger farmers. The consequences have been sharply polarized. On the one hand, smallholder rubber and rice production has steadily increased, with Malaysia now being 75 per cent self-sufficient in rice. On the other hand, poverty remains widespread. Between 40 per cent and 60 per cent of smallholder rubber producers and up to two-thirds of the rice producers are still in poverty. The proportion is falling but it is falling slowly.

The outcome of such trends in the rural areas has been widespread migration to the cities of Malaysia where industrialization has been enthusiastically promoted and where expansion has been spectacular (Plate 6.3). The key tactic of economic growth has been the development of industrial estates and free trade zones where the government has packaged an array of subsidies and concessions for the benefit of domestic and overseas investors. Manufacturing growth in Malaysia has undoubtedly been rapid (Table 6.3). But in addition to being spatially concentrated in certain zones in the larger cities, it has also occurred within a relatively narrow range of industries. Electronics alone accounts for over 40 per cent of the total growth.

Many of the industries are capital intensive so that their labour demands are relatively limited and tend to be for unskilled workers. This is particularly true for the electronics industry which mainly comprises the assembly of imported components with few backward or

Plate 6.3 Penang, Malaysia: industrial expansion
Source: J. Eyre

forward linkages. Moreover, wages in these leading industries are lower than the national average. The tendency to employ young women, who are traditionally less demanding, and the long suppression of trade unions have worsened this situation. The limited availability of formal urban employment in general, and of reasonably waged jobs in particular, has clearly been a factor in the *increase* in urban poverty since 1970. What has occurred, therefore, is simply a geographical shift of poor Malays into the cities.

The government might console itself with the fact that control of industry seems to be moving more towards the objectives of the NEP (Figure 6.3). Foreign capital appears to have declined in the face of growing Malaysian, particularly Malay, ownership. But non-Malay involvement is still double that of Malays and within the Malay sector it was institutional ownership that predominated. Indeed, with government encouragement and involvement, the period to the mid-1980s has been marked by increasing concentration of ownership of corporate assets.

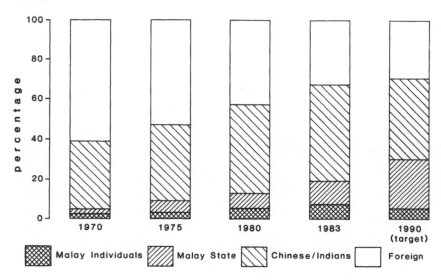

Figure 6.3 Malaysia: trends in corporate ownership

This overwhelming government role has, in fact, been the most consistent effect of the NEP and along with it has come a huge enlargement of the federal government bureaucracy which increased almost fourfold between 1970 and the mid-1980s. Needless to say, most of this increase has been Malay oriented – a process underpinned by government policy within the educational sector. Two-thirds of all university students are supported by government scholarships, 80 per cent of which are awarded to Malays.

When translated into household income data (Table 6.2), Malays have indeed caught up to some extent with the other races in Malaysia since 1970. But Malay incomes are still much lower on average because the gains have been narrowly concentrated into a small, largely bureaucratic, elite. Poverty is still widespread amongst both rural and urban Malays, indeed inequality amongst Malays has intensified since 1970, as it has amongst all other groups.

The impact of national policies since 1970 has, therefore, been contradictory, despite the apparently straightforward goals of the NEP. Essentially there has been a class-biased response with the middle- and upper-income groups of all races benefiting at the expense of both the rural and urban poor.

Conclusion: present and future problems for Malaysia

Two fundamental problems have resulted from the development strategies of the NEP. One is linked to domestic inequalities, the other is related to the extent to which Malaysian growth is reliant upon the world economy for funds and markets.

Of the two, the latter has elicited most response from the government, for by the mid-1980s the world recession had hit countries like Malaysia very hard. Not as resilient as the region's high-flyers, such as Taiwan or South Korea with their own substantive body of private capital, growth rates in Malaysia fell to near zero. This slump has been across the board with a catastrophic downturn in demand for (and price of) oil, oil palm, rubber, tin, and electronic components. This slump hit particularly hard at the recently emerged, state-sponsored, heavy-industrial sector (steel, cement, etc.) which is now more expensive than imported equivalents and cannot find sufficient domestic or international markets.

One dramatic response from the government has been to encourage privatization to recoup losses. This clearly goes against the spirit of the NEP since Malay jobs have been lost whilst foreign and Chinese investment shares have increased. In some ways this privatization has also been encouraged by widespread reports of corruption within state agencies and institutions, leading to serious schisms in the political establishment: something which does not increase international confidence in Malaysia.

Nor does it improve the internal political climate which has once again begun to sizzle, not this time because of ethnic dissatisfaction at unequal development but because of class conflicts. The urban proletariat is now more mixed than ever before, both in terms of ethnicity and gender, and chafes at its low wages and restricted bargaining power. Even larger is the urban informal or petty commodity sector, those unable to find waged work in the city and increasingly resentful of a *nouveau riche* riven by corruption and scandal. What is desperately needed in Malaysia is a political alliance between the urban and rural poor to seek fairer management of a potentially prosperous economy. Now that there are more Malays in the cities this may not be too far away. The alternative may well be a shift to Islamic fundamentalism.

Key ideas

1 Ethnic mixing characterized parts of Pacific Asia prior to the arrival of the Europeans, but colonialism accelerated and complicated this process.
2 Colonial labour demands were particularly important in changing the ethnic composition of the region.
3 The legacy of this ethnic intermixing has been bitter rivalry within the contemporary development process.
4 Such problems are present in both socialist and capitalist states.

7
Industrialization and the four little tigers

The dimensions of industrial growth: Pacific Asia in a global perspective

Over the last two or three decades, industrialization has proceeded very slowly in the Third World as a whole. The contribution of manufacturing to GDP is around 18 per cent, only a couple of points higher than it was in 1960. As a result the world export trade in manufactured goods is still dominated (about 80 per cent) by developed countries.

However, aggregate data can be very misleading and the Third World exhibits considerable diversity in manufacturing. One must be careful, in interpreting regional and national data to distinguish between overall levels of production and relative importance in world trade. Indicators of the former reveal that the largest producers are those countries with huge populations and, therefore, the biggest domestic markets. Thus China's manufacturing output is double that of the second nation, Brazil, whilst countries such as Singapore rank much further down the list.

On the other hand, indicators of manufactured exports show a different picture, with Pacific Asian countries dominating (Figure 7.1). South Korea, Hong Kong, Singapore and Taiwan are particularly important in this respect, with the first two alone accounting for a large proportion of the Third World's manufactured exports. When the data are calculated on a per-capita basis, Hong Kong and Singapore are world leaders in terms of manufactured exports, being 50 per cent greater than West Germany, the principal developed nation.

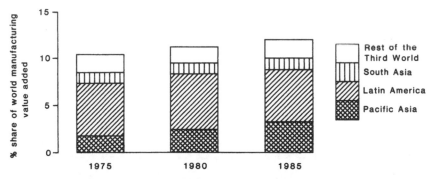

Figure 7.1 Manufacturing value added by major Third World regions

It is data such as these that have singled out the newly industrializing countries (NICs) of Pacific Asia, the so-called four little tigers (or dragons), for particular attention, especially as their impressive manu-facturing growth has occurred over such a short period of time. In the 1950s, manufacturing accounted only for around 12 per cent of GDP in each, but by the 1980s it was comparable to that of the most advanced industrial nations. Furthermore, there appears now to be a second wave of manufacturing nations rising in the Pacific Asia region, with Malaysia and Thailand being singled out by their rapid rates of economic growth.

However, not all of the four little tigers have exactly the same background to their success. There are several common denominators, some of which are local in origin, some of which are global, but the particular mix is unique to each state and the end-product slightly different. In our conclusions we will not only review the economic achievements of recent growth but also examine some of the social consequences.

Pacific Asia's four little tigers

1950s to 1970s

One of the principal common factors linking Hong Kong, Singapore, South Korea and Taiwan is that all four were colonies: the first two of Britain and the last two of Japan. However, there were considerable differences in the extent of local capital development, with Japanese colonialism laying a dead-hand on Taiwan and Korea in this respect, whereas in Hong Kong's and Singapore's domestic capital interests were well-developed in local retail and commerce.

After the Second World War a variety of factors combined to induce rapid early growth. Hong Kong and Taiwan, for example, received enormous refugee populations following the communist assumption of control in China. Most of these were urbanized and brought labour skills or capital to invest. To a certain extent this happened in South Korea, following the partition after the civil war, but not to any great extent in Singapore.

Also during the 1950s, the United States sought to develop Taiwan and South Korea as bastions of successful capitalism in an unstable region where communism seemed to be spreading rapidly. Much of this US aid and investment was in projects such as road and rail networks or education. Indeed, in Taiwan almost three-quarters of all infrastructural investment came from United States aid.

By the 1960s all four tigers had moved into export-based manufacturing. This was forced on Hong Kong and Singapore because of the decline in their regional trading roles and because of small domestic markets, but in South Korea and Taiwan the development was a natural extension of the already extensive overseas trade with the United States. However, entering a competitive global market is not so easy and there was a need to specialize in those areas of manufacturing that did not require extensive capital to produce a commodity of acceptable standard, i.e. clothing and textiles.

For Singapore, the circumstances were different since the republic had very limited sources of domestic capital, and the majority of its manufacturing has always been financed from overseas. Moreover, the range of industries was quite different, revolving around oil refining, shipbuilding and repairing, as well as manufacturing. But even in 1970 well over half of all exports were produced by foreign firms.

In many ways, the four economies expanded at the most appropriate time because the 1960s were marked by a consumer boom in the west which rapidly expanded the market for manufactured goods. In addition, the late 1960s and early 1970s witnessed the start of the outflow of MNC investment from Europe, USA and Japan.

As their economies expanded in the 1970s, so the four governments reinvested their expanding revenues in education, training and industrial services in general, all of which resulted in greater productivity and further encouraged domestic and foreign investment. Gradually production broadened to include more high-technology industries such as electronics. Meanwhile labour costs were kept relatively low by weak trade unions, strong government controls and, in the case of Singapore and Hong Kong, the use of immigrant labour (Malaysian and Chinese respectively).

In all four areas there was a steady increase in foreign capital, especially in Singapore where MNCs accounted for 80 per cent of total investment and two-thirds of manufacturing exports. United States and Japanese capital is particularly dominant, as is the resource on their export markets. However, in the 1980s this steady, if dependent, expansion was interrupted by the world recession. How did the four little tigers cope?

1980s: recession and recovery

The threat to continued high rates of growth in the 1980s came from several directions. They affected each of the states in different ways and prompted varied reactions. The first challenge came from the rising wage rates in comparison to other Asian or Latin American countries. For MNCs in which labour costs are important, this has led to a switch in emphasis, particularly to Thailand where rates of industrial growth are currently the highest in the region.

A second threat which has resulted from the world recession has been the consequent downturn in demand and investment from the narrow range of developed countries on which the four tigers depend. Associated protectionism has also led to a downturn in technology transfer from the west. In theory, Japan should be the obvious nation to step into the breach as far as markets and capital investment are concerned. Whilst there has been a sharp increase in the number of Japanese production plants in Pacific Asia, the bulk of the US$33 billion invested overseas has gone to Europe and the United States. Furthermore, the Japanese market remains closed to most Asian manufacturers.

Other factors which have posed problems for the four little tigers have been their very strong currencies, which have raised the price of export commodities, and the downturn in oil and commodity prices, which reduced the purchasing power of OPEC nations and the various regional markets in Asia. However, lower oil and material costs have also helped to lower production costs.

The impact of these difficulties has been varied. Singapore has been worst affected because of its dependence on regional markets (almost half of all exports), and the fact that these countries are now its principal rivals and seeking to protect their own fledgling industries. Moreover, the very high proportion of foreign investment in Singapore resulted in a substantive loss of production when rising wages frightened off producers in the early 1980s. Add to this the slump in oil refining and US markets for manufactured goods and the result was zero growth by the mid-1980s.

South Korea, Taiwan and Hong Kong have all been affected by their dependence on the US market (between 40 per cent and 50 per cent). South Korea, in addition, is one of the largest debtor nations in the world and must meet massive repayment rates each year; whilst Hong Kong and South Korea also suffered from political uncertainties or turmoil during the first half of the 1980s.

The response to this varied situation has been quite different in each state, despite their apparent similarity of economic activity. However, all have sought to stimulate domestic demand for industrial products as a means of overcoming the economic downturn. This has resulted in major urban construction projects, such as rapid transit systems, new airports, or new national rail and motorway links (in Taiwan and South Korea). But these are essentially short-term measures and in all the capital cities the construction boom appears to have peaked.

Individually, *Taiwan* has probably taken fewer measures than the other states. Diversification into high-technology and heavy industries, particularly shipbuilding, has been successful, but by and large Taiwan has simply ridden out the storm. In this respect, political stability and absence of a large foreign debt has undoubtedly helped.

South Korea, in contrast, has been more urgently in need of special measures because its high-debt service ratio means that it has to keep up its exports in order to meet repayments. Moreover, the inherent political instability within South Korea means that it cannot afford a lengthy recession and rising unemployment. Already trade unions are beginning to flex their muscles and the frequent use of strikes has both reduced production and pushed up wages (labour costs rose by 17 per cent in 1987). The result has been a deceleration in the growth rate to around 7 per cent p.a. in 1989. No doubt many countries would welcome such a growth rate but in South Korea it is causing apprehension.

The South Korean government has taken a major role in strengthening the economy, largely through its control of the five main banks. Diversification into heavy industries, such as iron and steel production, automobiles, shipbuilding and chemicals, has occurred with single-minded determination. For example, Ulsan City was a village less than

20 years ago; it now has 600,000 residents, half of whom are the families of Hyundai's car plant and shipyard workers.

Critics of South Korea's policies claim that its investment switch to heavy 'sunset' industries will not bring long-term economic security, particularly as dependence on foreign expertise is still quite high (Hyundai on Mitsubishi, Daiwoo on General Motors). Thus, although South Korea is the world's leading shipbuilding nation, profits are low and debts high. Certainly the government has been very slow to support diversification by small-scale, high-technology industries which currently account for two-thirds of employed workers in Taiwan and 72 per cent in Japan. Whether South Korea's politically inept but economically shrewd government can conjure up another successful development switch remains to be seen.

Singapore foresaw the world recession in the late 1970s and planned ahead for a second industrial revolution into high-technology industries. The government encouraged wage rises of 10 per cent p.a. in order to persuade manufacturers to shift to capital-intensive rather than labour-intensive forms of production. Singapore also hopes to become a regional centre for producer services such as banking, insurance and research. To cover the manufacturing decline in the short term, the government encouraged a massive construction boom in housing, hotels and the like whilst awaiting the transformation to the 'Switzerland of Asia'.

Unfortunately, the second industrial revolution had not occurred by the time the steam ran out of the construction industry. Singapore's labour costs are now higher than any of its regional competitors, discouraging foreign investors and resulting in the loss of down-market industries that were still proving to be profitable elsewhere. Moreover, competition from Sydney, Tokyo and Hong Kong slowed down the development of producer services. The result, as noted earlier, was two years of near zero growth in the mid-1980s and a strong reaction from the government. Wages were frozen and a substantial new package of subsidies was offered to foreign investors. A new burst of construction followed and soon GNP per capita growth was back into double digits. However, this still looks less secure than ten years ago and is far too heavily reliant on foreign funds. There has also been a social price to pay; a feature that will be examined in the conclusion to this chapter.

Hong Kong has probably proven to be the most resilient of the four little tigers, despite its uncertain future, without seeking to induce radical changes. The territory has retained its traditional (and profitable) industries such as garments, shoes and plastics but has also diversified into high-technology production too. Here the usual, small size of firms has been a distinct advantage. Hong Kong has also found it easy to increase its regional producer-services role, a development which has undoubtedly been facilitated by its position in relation to China. Many firms are anxious to establish a base in what will become part of a huge domestic market in 1997 (Plate 7.1).

Plate 7.1 Hong Kong: central business district. The massive office blocks reflect the popularity of the territory with multi-national firms anxious to establish a base in the future Chinese market. All of these buildings stand on land reclaimed from the sea.

However, Hong Kong has also capitalized on its proximity to China in more direct ways. Food and other basic needs are imported from China at stable, low costs, thus keeping inflation and wage demands down. Furthermore, the rapid expansion of the Shenzhen Special Export Zone (SEZ) immediately over the border has enabled Hong Kong manufacturers to fragment their production process and use the cheaper labour in China for extended, partial processing, without the necessity to invest in additional factory space. For example, it is estimated that 50 per cent of all production for Hong Kong toy manufacturers is done in China; Hong Kong undertakes only the research, design and packaging. Little wonder that 80 per cent of the foreign investment in China's SEZs originates in Hong Kong.

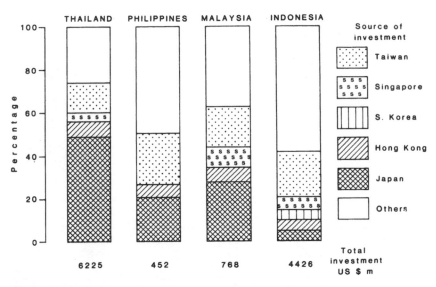

Figure 7.2 Four little tigers: investment in Pacific Asia 1989
Source: Far Eastern Economic Review

Such outward flow of investment is now a feature of all four little tigers. For Singapore, this primarily involves investment portfolios, particularly in US real estate, but for the other three the target is either resource production elsewhere in Asia, or MNC branch plants in the newly emerging industrial nations of the region (see Figure 7.2). More recently, Asian MNCs have sought to establish bases in Europe prior to full EC integration in 1992 when it is feared that external tariff barriers will be raised.

The recent growth in the outward movement of capital from the four little tigers indicates the extent and strength of their economy over the last two decades; a feature clearly confirmed by their resilience in the face of the world recession. It is, however, more difficult to put together a set of crucial common denominators that would constitute a Pacific Asia 'model for development' that so many other nations have sought to emulate.

A Pacific Asian model for industrial development

One thing is clear from any examination of Pacific Asia's four little tigers: their rapid development has not been due to purely economic reasons: political and social factors have also been highly influential. Clearly, however, economic explanations must underlie economic success and, in this context, one of the key elements has been the ability of all four states to mobilize and attract capital investment (despite their uncertain political circumstances). As noted above, Taiwan and South Korea received their initial capital boost in the form of US aid, whilst Hong Kong and Taiwan also benefited from the influx of refugee capital. Since the 1950s, however, all four have attracted high rates of both domestic and foreign investment. This has been fuelled by export successes and the prospects of high rates of return.

All four states also entered into world export markets at a propitious time when a consumer boom was in full swing, when MNC investors were looking for opportunities to expand production, and when the United States was keen to protect and foster successful examples of capitalism in a region thought to be at great risk of turning communist. All four, therefore, clearly benefited from aggressive US political activity in Pacific Asia.

Another consistent feature has been the central and obtrusive role of the state in acting on behalf of capitalism to facilitate its development. Such actions range from the creation of export processing zones, to the manipulation of state finances through the banking system, to shaping the capital cities by influencing the planning process (for example, by fuelling construction booms during economic downturns). This important governmental role is common to all four states, even Hong Kong, and strongly refutes the image of these economies as examples of freewheeling, *laissez-faire* capitalism.

It is also alleged that cultural traits, such as diligence, thrift or deference to authority, which relate to the mutual obligations and duties of Confucianism, have also been influential in inducing a work ethic into society. These characteristics operate not only within the household but at all other social levels too, creating loyalties towards the company and the state.

However, none of these adds up to a model for other countries to follow. It would be impossible, for example, to re-create the favourable world trade circumstances of the 1960s, or to transfer Confucian ethics to other parts of the Third World. Indeed, the Latin American NICs are quite different in their development process.

Perhaps the appropriate conclusion to draw is simply that industrialization has and continues to occur in Pacific Asia; that it is not necessarily a dependent industrialization (and is therefore of positive value to the state concerned); that industrial growth can be pursued in a variety of ways; and that the conditions for further expansion seem to be more established than ever before (as evidenced by Guangdong's attempts to become the fifth little tiger – see Case study L). But national economic growth is one thing, resultant societal development is another. How have individuals benefited from economic growth in Pacific Asia?

Case study L

Guangdong: Pacific Asia's fifth tiger

The Chinese province of Guangdong has ambitions to be the fifth of Asia's economic tigers and has moved some way towards this objective. The province contains 65 million people, as many as in the four other tigers, and its per-capita income is already 30 per cent higher than that of China as a whole. Moreover, local economists project that Guangdong's per capita income will be some US$4,000 by the next century, four times greater than the national figure.

The key to Guangdong's success has been its proximity to Hong Kong, from which comes 90 per cent of the province's foreign investment and to which goes 60 per cent of all exports. It follows that the area of most rapid expansion is in the Pearl River delta adjacent to the colony. This area contains two of China's four Special Economic Zones (areas designated for foreign capital investment), one of which at Shenzhen dominates the others, producing over 90 per cent of all SEZ exports.

Case study L *(continued)*

During the 1980s, however, with the growing liberalization of China's economic policies, other areas of production have emerged, notably Guangdong's own 'four little urban tigers' scattered around the delta region (Figure L.1). In addition, rural-based industries have also boomed, and overall the annual growth of industrial production has averaged at well over 25 per cent. The importance of rural-based production must not be underestimated: it currently accounts for 40 per cent of all Guangdong's exports.

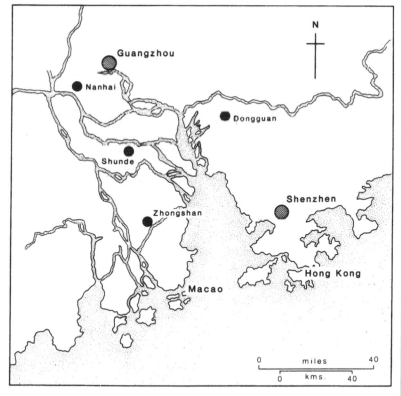

Figure L.1 Pearl River Delta: main urban areas

Case study L *(continued)*

The impetus for all of this growth has been processing work contracted from Hong Kong manufacturers, but this in its turn has brought about other advantages. The growth of job opportunities has attracted ambitious and skilled workers from all over China (most of whom have moved illegally to a province where incomes are undoubtedly higher. In addition, proximity to Hong Kong has increased the flow of foreign currency, which is a scarce resource within China. It is estimated that almost half of the black-market currency in the country originates in Guangdong, most of whose residents seem to have relations in Hong Kong. As a result, consumer goods are more readily available and cheaper in the province, acting as another incentive for migration.

Guangdong aims to quadruple its 1980 economic output by 1995, well ahead of the rest of China, and although it is still not as important in absolute terms as Shanghai or the northwest, the province leads the country in export earnings per capita. But in order to retain its pre-eminence, it has had to scour China for raw material and energy resources, outbidding less wealthy provinces and even investing in resource development projects in capital-scarce provinces. Needless to say, within the 'private market' the higher prices that prevail in Guangdong have induced flows of foodstuffs and other consumer goods, leaving neighbouring provinces complaining of shortages, even of the commodities they themselves produce.

Over the last year or two the Chinese government has tried to slow down Guangdong's headlong gallop, but provincial capitalists and administrators have the bit between their teeth and are likely only to pay lip-service to directives from Beijing. Some accuse Guangdong of moving towards an old-fashioned nineteenth-century capitalism involving considerable labour exploitation (including children), shady financial dealings and rampant consumerism. If so, then given the underlying Confucianist work ethic, the province seems well on its way to becoming one of the new-wave NICs of Pacific Asia.

Social change and economic growth

Although industrialization has contributed enormously to national economic growth, its impact on employment has not been proportionally as great. Although economic growth creates new jobs, Figure 7.3 reveals clearly that compared to its overall contribution to GDP, industry (which comprises manufacturing and extractive activities) has fallen somewhat short of expectations. What this means in real terms is that many of those who migrate to the city in search of factory employment may well find themselves disappointed and forced to seek work in what is commonly called the informal sector, which comprises a

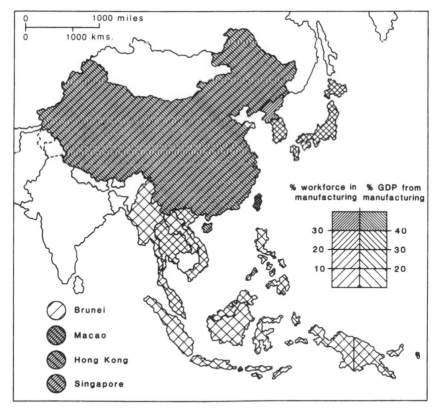

Figure 7.3 Pacific Asia: the contribution of industrialization to GDP and employment

myriad of small-scale, semi-legal or illegal activities. It must not be thought, however, that the informal sector is not linked to more formal, capitalistic enterprises; as Case study M indicates, the relationships are considerable and serve to facilitate manufacturing production.

Case study M

Domestic workers in the Philippines garment industries

The informal or petty commodity sector of the economy is difficult to define precisely, although many have tried. In fact, an enormous variety of non-conventional activities can be classified in this way, ranging from building a squatter house, through hawking cooked food on the streets, thieving, running a pirate minibus, to organizing a gambling den. The great majority of activities relate to providing retailing or other services, but a substantial manufacturing sector also exists.

Whilst many of the features outlined in Table M.1 do characterize petty-commodity manufacturing, there is a fundamental difference in the real world. The two sectors (whatever they are termed: formal/informal or capital/petty commodity) are not separate entities but are intertwined in a complex relationship in which one is dependent and one is dominant. In other words, those in the informal sector are exploited by individuals, enterprises and institutions in the formal sector. An excellent illustration of this can be found in Rosalinda Pineda-Ofreneo's account of outworking in the Philippine garment industry.

One of the fastest-growing industries in the Philippines, more than half of the garment industry is controlled by foreign firms, almost 30 per cent by US firms alone. The industry employs about 250,000 full-time factory workers, together with another half-million or so who work at home. About one-quarter of those who work at home are directly employed by firms but most work on a contractual basis, fulfilling orders placed with them by agents. The homeworkers are provided with materials (and sometimes machines) and are paid when the goods are delivered to the firms. More than 90 per cent of these workers are women, and their wage rates are only one-sixth of those in Japan.

Case study M *(continued)*

Table M.1 Differences between the formal and informal sectors

Informal sector	Formal sector
Ease of entry	Difficult entry
Indigenous inputs predominate	Overseas inputs
Family property predominates	Corporate property
Small scale of activity	Large scale of activity
Labour intensive	Capital intensive
Adapted technology	Imported technology
Skills from outside school system	Formally acquired (often expatriate) skills
Unregulated/competitive market	Protected markets (e.g. tariffs, quotas, licensing arrangements)

Source: International Labour Office

The homeworkers are usually not covered by labour laws, including those relating to minimum wage legislation, and working conditions often contravene safety regulations. These informal sector workers are, therefore, cheaper than factory employees and can be used to undermine wage demands from the formal work-force. Moreover, firms can easily overcome expansion and contraction in demand by use of subcontracted outworkers without the company needing to adjust its plant or its formal labour force.

The system extends throughout the urban hierarchy and into the rural areas where rates of pay are lowest. Rosalinda Pineda-Ofreneo reports that women working all day and sewing babywear earn just over US$1.00, from which is subtracted the cost of thread as well as transport to and from the agent who commissions the work. In real terms, moreover, the pay has declined over the years as inflation has accelerated.

The worst aspect of the whole system, however, is that the female outworkers are not only exploited themselves but are used to prevent their sisters in factory employment from unionizing to obtain better working conditions. When such moves threaten, the firms simply dismiss the leaders and reallocate more work to subcontractors.

Despite some misgivings about the extent of job creation, many observers feel that the growth of manufacturing in the four states has brought broadly based benefits to their societies. In material terms, low-income households are probably better-off than their counterparts in other Asian countries. Real wages rose at rates of between 5 per cent to 10 per cent until the 1980s when the world recession reduced the demand for labour. Moreover, some would claim that the distribution of income has also improved, although not in an entirely satisfactory way. Although the total share of the top 20 per cent of income earners has everywhere declined, gains have been made by the middle classes, so that the lowest 20 per cent of income earners in Hong Kong and South Korea, for example, still only earn around 5 per cent of total household income; very similar to the pre-expansion figures of the 1950s and comparable with the current average for the Third World as a whole. Moreover, as the middle class has adopted western life-styles, so wide gulfs have opened up between the haves and have-nots (Figure 7.4).

But the unequal distribution of income is not the only social cost of

Figure 7.4 Taiwan: rich and poor
Source: Far Eastern Economic Review

economic growth. In each of the four tigers, state intervention has tried to ensure that its labour force remains cooperative and does not press for the wage demands that may scare away potential investors. Thus, trade unions tend to be somewhat toothless industrial peace-keepers rather than defenders of workers' rights. Indeed, in these admired bastions of capitalism, democratic rights for the individual tend to be rather patchily observed. Whilst ostensibly open elections are held from time to time, gerrymandering occurs in South Korea and Singapore (where the ruling PAP has two-thirds of the vote but 95 per cent of the MPs), whilst many seats in Taiwan are still reserved for the ruling Kuomintang. Hong Kong, of course, does not yet have universal suffrage for elections nor a representative government.

Over and above these aspects of the political scene, many have claimed that individual human and social rights have also been threatened in the four little tigers. In Singapore, for example, a big-brother situation allegedly exists whereby children are rigidly streamed from primary school onwards into 'hand' or 'brain' groups. To the visitor, Singapore appears to be a city of signs exhorting people to behave correctly or else suffer the consequences, usually a $500 fine, but most of its inhabitants clearly interpret the situation in a more benign way, feeling political and economic security to be worth a few personal sacrifices.

Sensitive to criticism of social engineering, some governments have not only controlled or banned domestic and international media, but have also threatened to disenfranchise those who use their vote 'unwisely'. Even worse have been the arrests without trial of members of various pressure groups seeking greater freedom of expression, and of the lawyers who try to defend them. All this is done using similar legislation to that employed against the leaders of current governments in the immediate post-war years. For too many people in Pacific Asia the social cost of economic growth is still unacceptably high.

Key ideas

1 Pacific Asia contains several of the most rapidly industrializing Third World countries, notably Hong Kong, Singapore, Taiwan and South Korea.
2 The industrial contribution to GDP is not necessarily matched by its importance in employment.

3 Political and cultural factors have been as important as economic factors in encouraging industrial growth.
4 Throughout Pacific Asia both state governments and external investment have played important roles in bringing about rapid industrialization.
5 Although there are common denominators, each of the regional high-flyers has its own industrial features.

8
Urbanization and urban planning in Hong Kong

Pacific Asia: the dimensions of urban population growth

By world standards, Pacific Asia is not highly urbanized. Just over one-third of its population is classified as urban, a much lower proportion than in Europe, the Americas or the Middle East. However, urban growth rates are high, being twice those of the Third World as a whole. The reason for this apparent contradiction is that rural populations are substantial and continue to grow rapidly throughout Pacific Asia. The consequence is both intensive rural and intensive urban development, with 10 of the world's 30 most populous cities located in the region. These figures suggest that urban growth tends to be concentrated in large cities and this is indeed the case, although with considerable regional variations. In fact, such diversity is typical of Pacific Asia, as Figure 8.1 reveals, making explanation of current trends somewhat difficult.

Patterns of any sort are difficult to discern within the region but one or two points are worthy of note. First, the most advanced economies, the four little tigers of Hong Kong, South Korea, Taiwan and South Korea, do emerge as having a higher urban proportion of their total population. Indeed, the East Asian states as a subregion can also be characterized in this way (inclusive of Japan). Conversely, those countries with the weakest economies are characterized by the most rapid urban growth. This is a feature common to the Third World as a whole and not only reflects a natural deceleration of urban growth once

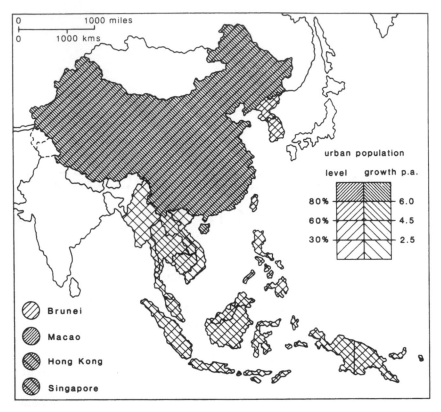

Figure 8.1 Pacific Asia: the extent and growth of urban populations

the majority of the population becomes urbanized, but also indicates the
way in which urban centres are perceived as growth points in poor
economies. These generalizations hold for both socialist and capitalist
states.

Primacy in the region seems to vary considerably. At one extreme is
one of the most primate cities in the world, i.e. Bangkok which is
50 times larger than the second city in Thailand; on the other hand
many states exhibit less primacy than many European countries,
averaging between 20 per cent and 30 per cent of the total urban
population.

The inconclusive nature of this review of urban population data for
Pacific Asia can, in a way, be quite instructive since playing with
numbers clearly only scratches the surface of urbanization in the region.

What we really need to know is why cities have grown. The simple answer would be because there are now more people in them. But the urbanization process is much more complex than this. People move to cities for a multiplicity of reasons and in response to circumstances over which they have little control. This has become particularly true in Pacific Asia over the last two or three decades as the way in which the region and its resources have been drawn into a changing world economy has become more complex.

Post-colonial trends in economic and urban growth

Since the 1950s, the urbanization process in Pacific Asia has been shaped by its increasing incorporation into the global economy. This was neither a smooth nor uniform process, varying with the basic resources of each country, their histories and political structures, as well as their particular relationships with various core nations.

In the 1950s and 1960s the economies of most countries in the region changed very little. They still relied heavily on primary exports but were trying to develop import-substitution industries in their cities. Lack of investment capital and limited internal markets (because of poverty) meant that in general, economic growth was sluggish, apart from those countries that were being directly supported by US aid. These included Japan, South Korea and Taiwan, which the United States envisaged as bastions of successful capitalism stemming the rising tide of communism in the region. The aid gave these countries a lead in urban industrial development which they still retain.

Notwithstanding the variation in the rate of economic growth and diversification, urban populations throughout the region expanded rapidly as large numbers of rural migrants moved to the cities in search of the jobs they felt that independence and development was bound to bring. However, such jobs were not forthcoming for the great majority who nevertheless stayed in the city eking out a living wherever and however they could. It was during this period that the basis of the large urban petty commodity (informal) sectors was established, often in large squatter settlements.

By 1970, however, a fundamental change had occurred in the world economy. Diminished economic profitability in the core countries of Europe and North America had induced a shift in capital investment into countries where production costs were lower and productivity per capita higher. Cheaper labour was an important factor in this programme of reinvestment, but other considerations were important too, such as an educated (and trainable) workforce, good port facilities, a degree of local capital (which usually provided the bulk of the investment funds), and political stability.

Satisfying these conditions was not easy, and Pacific Asian countries began to be incorporated into this process of investment in a somewhat erratic way. South Korea and Taiwan were already industrializing, whilst Hong Kong and Singapore provided ready pools of skilled labour and Chinese capital, but elsewhere the political situation seemed rather unsettled and only unusual resource access attracted overseas capital.

With the parallel technological revolution in communications of the post-1960s – from containerization, telexes, telephones, computers through to fax machines and satellites – the production process became fragmented. The labour-intensive parts shifted to the Third World in what became known as a new international division of labour.

This process of industrialization is clearly urban oriented. Those countries which were not the immediate targets for investment sought to make themselves more attractive by development packages offering a range of financial inducements to investors – tax concessions, free land and so on. Most of these concessions were offered in self-contained spatial packages; free trade or export-processing zones located in or around the capital cities.

As industrialization expanded erratically around the region, it further induced population movement into the cities. But these population shifts were different; women were in greater demand than men in the new multinational factories and began to dominate the migration process within certain age groups (see Chapter 9 for a full discussion). As industrial expansion occurred, other economic and social changes began to affect Pacific Asian cities in their wake. Expanding corporations producing for a global market require sophisticated back-up facilities (financial, technical, training, etc.), so that a range of producer services began to appear, again quite varied in their intensity and level throughout the region. Moreover, as urban incomes rose, so did the demand for consumer goods, including those provided by the state (housing, schools, hospitals). Given the international linkages of these cities, it is not surprising that many of the consumer services were western in character, aping and adopting the values prevalent in New York, Bonn or Rochdale.

In short, the whole nature of the urbanization process has both broadened and deepened but enormous variation exists within and between the cities of Pacific Asia, depending on the extent to which they have been incorporated into the capitalist world economy. Clearly there will be massive differences within the urban hierarchies of individual countries; international values and wealth tends not to penetrate too far downwards, making life in a small provincial town far removed from that of the capital city. But between capital cities in Pacific Asia there are also massive contrasts, as we have already noted. Life for even the wealthy in Rangoon or Pnomh Penh, for example, would be very unsophisticated compared to that of a lower-middle-class household in Singapore or Hong Kong where colour televisions, washing machines and other consumer goods are the norm rather than the exception. However, whatever the particular material circumstances, cultural, political or economic, almost all cities in the region are growing fast and their physical environment has had to change accordingly.

In some cities this physical change in the built environment seems unsupervised and chaotic, but most have tried to control, and if possible, plan for growth and development. This is where conflict arises between economic and social demands. Is the city to be shaped by considerations of efficiency or equity? Indeed, need the choice be so extreme and can the state successfully intervene in collective consumption in order to balance the response to these demands? It is such questions which will comprise the thematic focus for the following examination of the urbanization process in Pacific Asia; the geographical setting for the investigation will be one of the most sophisticated cities in the region, Hong Kong.

The evolution of contemporary Hong Kong

Hong Kong is a compact city-state of just over 1,000 square kilometres situated like a pimple on the backside of China. Politically, it was acquired in three stages. Hong Kong Island was ceded to Britain in 1842, the Kowloon peninsula in 1860, whilst the New Territories were leased for 99 years in 1898 (Figure 8.2). The physical development of what is still a colony reflects these divisions, but all are due to be returned to China in 1997, after which Hong Kong will become a Special Administrative Region in which its existing economic system will, in theory, be allowed to continue.

There is no representative government in Hong Kong, nor is one anticipated after 1997. The colony is effectively governed by its civil service, through 40 government departments which employ some

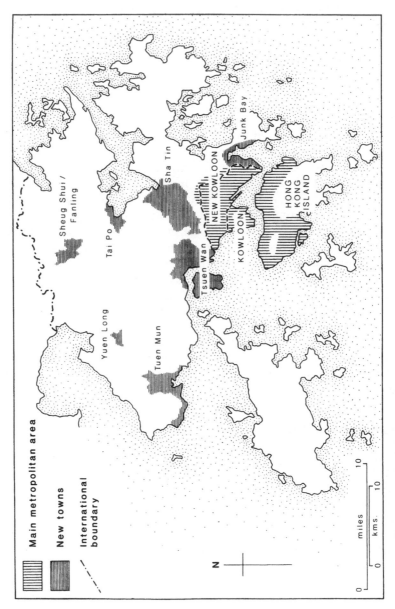

Figure 8.2 Hong Kong: main urban areas

200,000 people. It is a common fallacy to think of Hong Kong as a *laissez-faire* economy in which 'anything goes'. As this chapter will indicate, it is, and always has been, a closely controlled and tightly administered political and economic entity.

For over a hundred years, until the 1950s, Hong Kong was primarily a port, through which most of China's trade passed. Urban development has concentrated around the harbour on the Kowloon peninsula and the northern shore of Hong Kong Island. Much of this area comprises steep-sided hills so that high-rise housing on the limited amount of flat land has, therefore, tended to be a tradition in the colony.

The economic and urban situations changed dramatically in the 1950s. Civil war in China and regional conflicts in Korea and Vietnam all ruined Hong Kong's entrepôt role and the colony began to invest in the manufacturing production for which it has subsequently become so well known (see Chapter 7). Much of the initial financial and human capital for this transformation came from the Chinese refugees who flooded into the colony, legally and illegally, until the 1960s (and subsequently in the 1980s). Later investments were, however, more representative of the wave of capital flows that gave rise to the new international division of labour in the 1970s.

Hong Kong was one of the earliest of the Asian NICs and, moreover, had experienced a considerable amount of domestic entrepreurial development. It was, therefore, not surprising to see the colony subsequently develop as an international centre for financial transactions and other producer services. Hong Kong is now the third-largest financial centre in the world, and yet still has a substantial manufacturing capacity which employs one-third of the workforce and makes the colony the world's leading producer on a per-capita basis. As noted in the previous chapter, Hong Kong's geographical position *vis-à-vis* China has been a crucial element in this continued economic growth, providing cheap food, cheap labour inputs and attracting international investors because of the prospect of access to a hugely expanded domestic market in 1997. At least that was the situation until the brutal repression of the Chinese democracy movement in 1989, an event which cannot fail to undermine confidence in the economic future of Hong Kong.

Most of this economic expansion has fuelled and been fuelled by an equally rapid growth in Hong Kong's population (Figure 8.3). This was the result of an erratic mixture of natural growth and immigration (both legal and illegal), and although the overall rate of increase has now slowed down to one of the lowest in urban Asia, the limited size of the

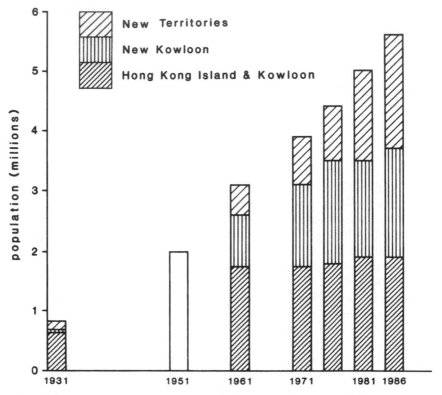

Figure 8.3 Hong Kong: population growth within the main urban areas

colony has resulted in extremely high population densities. In 1987 the World Resources Institute named Hong Kong as the most densely populated city in the world, with almost 105,000 persons per square kilometre in the urban residential areas, over two and a half times the figure for Lagos, the second city on the list.

Clearly, intense population pressure and rapid economic change are going to have an enormous impact on the built environment and in a small place like Hong Kong planning for urban growth must have a high priority otherwise chaos will ensue. The crucial question is what are the goals of such planning? One must be to create an economically efficient city, but does this mean that social objectives or environmental considerations take subsidiary roles?

This distinction is very evident if the evolution of the colony's housing programme is examined. I have discussed this at some length elsewhere in this series and there is little point in repetition here, particularly as some related issues will be examined later in this chapter in the context of regional planning. However, it is worth noting that all major changes of direction in housing policy in Hong Kong – the decision to opt for a mass low-cost programme in 1954, the switch to higher standards in 1963, the shift into an extended new towns programme in the 1970s, and the introduction of home-ownership options were all made at the highest levels of political management. The planning and housing departments simply got on with their programmes within the guidelines set for them.

Indeed, it is important to appreciate that in recent years the urban managers of Hong Kong have realized not only that the built environment can be manipulated by the 'neutral' planning process for a variety of political, economic or, more rarely, social ends, but also that it can become a vehicle for substantial profit and gain in its own right. All land in Hong Kong is owned by the Crown but leases, as they become available, are auctioned to the highest bidder and realize millions of dollars for the Hong Kong government. New leasable land is constantly being created in Hong Kong by reclamation of harbour areas or levelling of mountains. In addition, huge profits are made out of the constant demolition and reconstruction of existing buildings to create ever-more space for offices, hotels, shops and so on. The incessant reshaping of the built environment in Hong Kong not only facilitates the expansion of capital but it also creates capital and wealth in its own right (Plate 7.1).

We are now some distance from the original starting-point of this section, i.e. that cities grow because more people move in. Clearly it is not as simple an equation as that. However, millions of ordinary people do live in Hong Kong and we must constantly ask the question of how these residents benefit from the way in which the built environment is reshaped by developers, planners and urban management.

In this context, Hong Kong furnishes an excellent focus for examination. Although it seems almost unique in its political geography (akin only to Singapore), it has become a model of sophisticated, successful urban growth to which most Asian capital cities aspire, at least those outside the socialist states: a place described by Peter Hall in *World Cities* as 'one of the most forward-looking great cities of the world'.

Clearly we cannot investigate all aspects of this urban development process and must be selective. Fortunately, as noted above, some discussion on housing is already available elsewhere in this series and to complement the coverage of this important basic need, we will first examine urban transportation and second, planning for growth in the immediate city-region. In this investigation, Hong Kong will provide the central empirical material, but considerable supporting evidence will be drawn from other cities in Pacific Asia.

Urban transportation

Changes in transportation over the last two decades have massively affected the cities of Pacific Asia. In particular, the improvement and cheapening of long-distance travel has brought many more people into the urban areas. Within the cities, transportation changes have been more complex, reflecting the role which mobility plays in urban development. Not only do the various modes of transport directly link producers and consumers, or workplace and residence, they also provide important employment opportunities in operating, maintenance, road construction and repairs, and so on. In Hong Kong, it is estimated that more than 100,000 people are directly employed in transportation and communications.

In short, transportation is a vital part of urban growth and it accordingly consumes a large slice of most urban planning budgets (usually between 15 per cent and 25 per cent). Yet the situation in Pacific Asian cities, with regard to transport, is very much the same as for other urban services – there are too many people and too few facilities. This problem is exemplified by the growth in automobiles which, in every city, has far outstripped even the rapid expansion of population. In Hong Kong, for example, between 1961 and 1981 the population grew by 53 per cent but the number of automobiles increased by almost 700 per cent. In other cities the situation is just as bad: Jakarta's automobile population increases by almost 10 per cent each year.

But this is not the complete picture, for in most cities the vast majority of the population do not own cars and do not find public transport either cheap or convenient. There are varied consequences of this. In some cities, it means that the poor will need to live very near to their (potential) places of work, thus giving rise to overcrowded slum and squatter districts fringing major sources of employment, such as docks, industrial zones and central business districts. With regard to transport itself, it means that the poor have invariably developed their own systems based on petty commodity (informal) services, such as illegal minibuses or motorized and non-motorized trishaws or cycles. Many more simply travel by foot, and surveys in Indonesia, Malaysian and Filipino cities indicate that between one-third and one-half of all journeys are pedestrian rather than vehicular.

When put together this adds up to enormous congestion as different modes of transport, pedestrian, animal, pedal or mechanical, compete for limited road space and all-day traffic jams are not uncommon in many capital cities. Even China is not free from this problem, with almost 7 million bicycles competing with almost half a million buses and trucks for inadequate road space. As with all aspects of urban planning, there has been a response from the state to this worsening situation, but in transport more than any other field this has highlighted the conflict between efficiency and equity in urban planning.

Changing responses

The planning responses to these transportation problems have been strongly influenced by prevailing views on development strategies as a whole. Thus the 1950s and 1960s were marked by an increasing tendency towards modernization along western lines. For urban transport this meant that small vehicles with flexible routes were replaced by larger vehicles with fixed itineraries. Small enterprises were amalgamated or integrated into larger, more bureaucratic firms, or else were squeezed out of business by new legislation prohibiting their operations at certain times of day, on particular routes or in specific areas of the city.

Most of the guidelines and much of the technology for transport planning in Pacific Asia came from the west via Japan. Japan itself had been reconstructed after the devastation of the war with US aid and along the principles of 1930s urban planning. In short, urban regeneration in Japan was structured around the automobile, in the shape of private cars and public buses. With the reassertion of Japanese political and economic independence from the United States in the 1960s, Japan began to develop its own peculiarly bureaucratic and formalized response to urban planning needs. Transport plans were, therefore, devised for settlements of varying sized and lodged in a massive procedural manual at the Ministry of Construction.

Gradually, as Japanese influence within Pacific Asia became more pronounced, particularly after Japan sought to improve its regional image through assistance programmes, these proved to be very 'exportable'. Teams of advisers would go into Asian cities of all sizes and make recommendations based upon Japanese experience and involving the transfer of Japanese technology and hardware. As a result, Japanese firms have acquired a sizeable slice of the urban transportation market in Pacific Asia; this means considerable profits. Thus much of Japan's aid programme is given in terms of material goods, such as its recent US$150 million loan to Beijing for two new subway lines for the 1990 Asian Games. International aid too offers opportunities in this respect. World Bank loans for transport in 1986 alone amounted to almost US$1 billion, more than one-quarter of all loans given to the region.

Case study N

Singapore: incorporation of the bus system

Prior to independence, Singapore's bus services comprised two distinct types. First, there was the British-managed Singapore Traction Company (STC) which dominated the main routes to and within the city centre; second there were a number of small companies operating small jeep-type vehicles known quite appropriately as mosquito buses that buzzed around in the spaces between STC's operations. There was little integration of these services with the planning process; new routes tended to lag behind suburban developments because of the need to apply for licences, with the initiatives coming from the companies themselves.

In 1959, self-government was granted to Singapore and its new Assembly had a strong socialist bias. Education, housing and employment were its priorities, and transportation received little attention, despite the fact that many large housing estates were beginning to appear in what was then the periphery of the city. The STC and the various Chinese bus companies had their recognized spheres of influence and found it difficult to agree on new routes to save the burgeoning suburbs. As the government would take no initiatives in the matter, the informal sector surged to respond and a massive pirate (illegal) taxi service sprang up. Between 1955 and 1962 the number of licensed taxis doubled to 3,200, but the number of illegal cars rose from 5,000 to 8,000. So

Case study N *(continued)*

flexible, so competitive and so cheap were these services that the bus companies, especially STC, began to lose money. This further reduced their capacity to improve their bus fleets and offer effective competition.

The Singapore government did not really turn its attention to transportation planning until the 1970s, following its review of the Master Plan and the decision to massively expand housing and economic development through a series of new and expanded towns located all over the island. By this time the chaotic, uncoordinated bus services and pirate taxis were regarded as 'not consistent with the Republic's aim of building a law-abiding, disciplined, tightly-knit and organized society'. The solution was to be modernization and incorporation of the bus services and ruthless suppression of the pirate taxis.

Measures against illegal taxis included suspension of driving licences, scrapping of cars and detention, even imprisonment, of drivers. Any resulting unemployment, the government felt, would provide extra manpower for public transport or the expanding factories. In 1970, the bus services were amalgamated into four major companies and the government began to initiate routes and invite tenders. Thus, for the first time, transport planning was integrated with other aspects of physical planning.

However, the separate bus companies proved not to be capable of modernizing and rationalizing to the extent that the government wishes. In 1973 they were merged into the single Singapore Bus Service (SBS), and the government drafted in officials (mostly from the armed forces!) to reorganize the management, even though it was, and still is, a private company.

The eventual success of this modernization and incorporation process soon induced neighbouring cities to follow Singapore's example. The Metro Manila Transit Corporation was established in 1974, the Bangkok Mass Transit Authority in 1975, whilst municipal authorities in several Indonesian cities also expanded their operations. However, without the close supervision and interference of the state, these ventures have proven to be unwieldy and unprofitable. Meanwhile, in Singapore, modernization

Case study N *(continued)*

and incorporation continues apace. The Mass Rapid Transit (MRT) subway opened its first line in 1982 and by 1990 was complete, with about half of the island's population living within 1 kilometre of the route. This, it is hoped, will finally squeeze out the independent minibus operators who still fill in the interstices of the SBS network, taking children to school and so on. Full integration of transportation within the state planning structure will then be complete. This even extends down to the behaviour of the passengers; travelling with a damaged ticket on the MRT can incur a fine of up to S$500!

Source: Based on a report by Andrew Spencer.

These modernization strategies have, however, enjoyed only limited success. Much of the apparently sophisticated analyses found it difficult to incorporate data on the complex, rapidly changing land use and transportation patterns of the region's cities. The inputs were therefore patchy and unreliable. In Manila, for example, the 1 per cent sample failed to capture the variety of modes of transport and land uses, ignoring squatter settlements and walking.

The results were, not surprisingly less than satisfactory. Incorporation meant less frequent and flexible services, longer waiting and travel times; moreover they reduced substantially the opportunity for employment in the transport industry (since many small enterprises were replaced by a few large ones). Even in those cities considered to be exemplary representatives of the modernization process, considerable difficulties were experienced and 'alternative' transport kept reappearing (see Case study N).

The reaction by many authorities was to suppress the competition from these 'alternatives', and non-motorized trishaws have been banned from all centres of Manila, Bangkok and Jakarta. Soon, however, research into informal sector activities revealed the positive contribution that they made to urban life. These views were reinforced by the many problems being experienced by the incorporation programmes. Heavy subsidies for road construction and bus companies were benefiting few urban residents, whilst the cleared road space was being absorbed by an expanding population of private cars.

Recently, therefore, many cities have sought to reflate their non-conventional modes of transport, most notably through the development of minibus systems operated by small companies (see Case study O). By encouraging such systems, state subsidies are reduced, free competition is fostered, and new technologies are promoted in the non-conventional sector (for example by promoting minibuses instead of pedicabs or trishaws). Needless to say, Japan was once again well placed to benefit from this switch in tactics.

Although non-conventional modes of transport have received considerable encouragement over the last decade, this has not stopped developments in the conventional transport sector. Indeed, the financial leeway provided by the self-sustaining activities of small operators has enabled the authorities to be even more ambitious in their transport planning – an ambition which is well represented in the urban metro or subway system.

Hong Kong was the first of the region's cities (outside Japan) to commit itself to a mass transit system in 1979. Currently Singapore, Seoul and Taipei are following suit, with Manila and Bangkok planning elevated urban rail networks. Each of these major undertakings is costing between US$5–7 million and has resulted in lucrative contracts for British, American and Japanese consultancy and construction firms. Inevitably, fares have had to match costs and, although subsidized, are usually beyond the pockets of the poor, resulting in mass transit for the middle class.

Hong Kong mirrors all of these general processes, albeit with a greater degree of success in promoting both efficiency and equality. Vehicle registration in Hong Kong has accelerated remarkably in recent years (Figure 8.4), although this does not reflect a high degree of car ownership. Household car ownership has only recently moved into double figures in the territory. This is not so much a consequence of an unequal distribution of income but rather a reflection of the inconveniences of private car travel compared to the benefits offered by public transport.

In common with other large cities in the region, Hong Kong's conventional transport system was nearing saturation point by the mid-1960s. Stimulated by the confusion caused when the 'culture revolution' spilled over into Hong Kong, thousands of illegal minibuses took to the streets and for a while gave rise to chaotic traffic conditions in the metropolitan areas. These were quickly legalized and by the late 1970s acceptable ceilings for the number of vehicles had been fixed with the

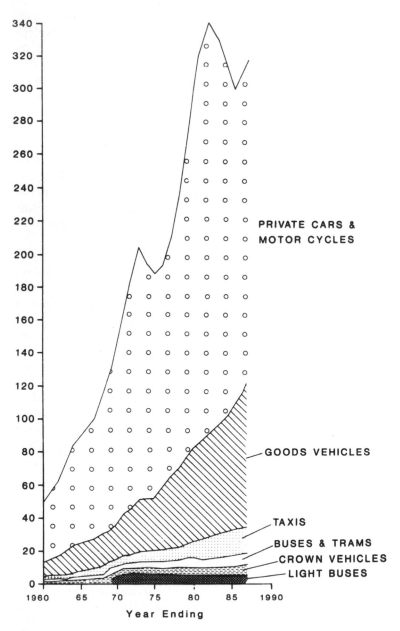

Figure 8.4 Hong Kong: vehicular growth rates

minibus service coming firmly under governmental control. As Figure 8.4 indicates, this mode of travel has become as important as the conventional modes which appear to have suffered little as a result of the competition.

Since the 1970s, Hong Kong has sought to be more comprehensive in its transport planning, rather than react to changing circumstances. This has resulted in major development projects which have taken the initiative in restructuring transport patterns. Examples include the cross-harbour tunnels, the mass transit system and, more recently, the new container terminal and airport. In addition, as the following section will show, macro planning strategies in Hong Kong have seen deliberate spatial emphasis given to the New Territories which, in itself, has resulted in new transport developments such as the electrification of the Kowloon–Canton railway and the introduction of a new light railway.

Despite the steady incorporation of transportation into the broader planning process, transport planning in Hong Kong 'continues to depend heavily on overseas consultants and to completely by-pass public participation' (C. K. Leung). In short, decisions are made more on economic grounds than any other.

Case study O

Kuala Lumpur: the rise of the minibus

Kuala Lumpur provides a classic example of a major switch in transportation policy. Prior to the 1970s transport planning in the Malaysian capital was firmly geared to the creation of a western-style city. The transport plan of 1963/4, for example, budgeted 98 per cent of its expenditure for new ring and arterial roads and only 2 per cent for miscellaneous services, including public transport. By 1970, however, priorities had shifted to getting the masses to work on time and a state-run bus service was introduced by means of legislation to combine the existing companies and the support of a World Bank subsidy.

In a parallel project typical of this era, another World Bank study recommended and supported construction of a six-mile, six-lane highway linking the industrial zones of the new town of Petaling Jaya with its labour force in Kuala Lumpur. As Petaling Jaya itself was a middle-class dormitory town for white-collar

Case study O *(continued)*

workers in Kuala Lumpur, the new highway filled a desperate need but can hardly be said to have been part of an overall development strategy. Moreover, facilities for movement within Kuala Lumpur itself, particularly to and from the poorer districts, were inadequate.

By the mid-1970s, only a few years later, the conventional wisdom had changed again and the values of small-scale operators were becoming appreciated. Accordingly, in 1975 yet another World Bank study recommended the introduction of minibus networks (there were already many illegal operators). Some 400 were originally licensed and this has now risen to 500 which carry about 13 per cent of all passengers in the city.

Some transport analysts have hailed this as a great success and, indeed, such favourable views strongly influenced Britain's own bus deregulation in 1986–7. However, a closer look at the system in Kuala Lumpur will reveal considerable state intervention to favour the minibuses. First, their fares are restricted to a flat rate which means that minibuses are cheaper to use on longer (and therefore more profitable) journeys from the suburbs (using standard bus stops); second, they do not need to comply with concessionary fare regulations, for example for students; finally, their licensed operators are forbidden to increase the size of the vehicle (even if they wished to). The consequence is a system which is quite the opposite from the short, feeder services to which minibuses are supposed to be most suited. Whilst the companies operating the minibuses are very small and therefore numerous, the system they provide constitutes a manipulated petty-commodity service which may make urban movement more efficient but is not geared towards the low-income groups in either a spatial or social sense.

Planning for the city-region

Background discussion

The problems posed by rapid city growth, particularly as a result of rural-to-urban migration, have attracted considerable attention since the 1960s. Few responses have been coordinated to any great extent.

As Chapter 4 indicated, strategies for promoting rural and regional development tend to confuse the sectoral and spatial dimensions of growth and have had little real impact, unless linked to strong state intervention at the village level. Even within 'successful' rural development programmes, there is little or no integration with a prepared strategy for urban growth. Very few countries have national urban development strategies and even fewer have national urban and regional development strategies.

For the most part, rural–urban articulation has occurred through the medium of urban decentralization schemes, of which the most popular was the growth-pole approach. This attempt to decentralize a miscellany of urban functions and services into the rural areas is basically a western strategy and has had little success in either rural or regional regeneration with the Third World.

And yet, despite the succession of failures of planned programmes of urban decentralization, the distinction between what is urban and what is rural has become decidedly blurred over the last few years. The reasons for this are not difficult to discern. Improvements in public transport have widened the commuting field of major cities, and also have led to increased circular migration. At the same time, improvements in communications via the spread of radios, television and telephones have increased knowledge of urban customs and values, as has the spread of educational services. These changes in transport and communications have also altered the employment structure of the regions surrounding many settlements. No longer are their residents reliant on agriculture; an increasing proportion work in non-farm occupations either in the city or in the rural areas themselves. This is not to say that the urbanization process has been fully diffused outside the major centres. The bulk of secondary or tertiary sector jobs, and such welfare services as exist, are heavily concentrated in the big cities. Indeed, even within the urban hierarchy itself, there are still wide contrasts in the quality of life.

What this suggests is that there needs to be more attention given to the question of planning for the city-region. It must be made clear at this point that we are *not* discussing regional planning which is an altogether broader issue and involves national development goals. The issue here is planning for the better integration of cities and their immediate hinterlands. In fact, it may well be the case that planning for the urban-region accentuates national regional imbalance by making the core areas seem even more attractive, leading to major problems, such as urban sprawl, chaotic transport patterns, or the loss of agricultural and recreational land. Such problems are already very evident in the peripheral areas of most cities in Pacific Asia where lack of interest and supervision has resulted in land (which is extremely valuable for low-income groups for subsistence agriculture, fuelwood or squatter housing) being taken over by speculators to lie idle and unused.

In many cities, where central area congestion is high, the peripheral sprawl of activities has been allowed to continue unchecked, although some authorities have attempted to divert rapid growth into satellite or new towns. Only in the restricted situations of Hong Kong and Singapore have efforts been made to plan for organized development in the region immediately surrounding the main metropolitan area. Even in these locations, efforts have been relatively recent, uncoordinated and had mixed success (either in terms of efficient or equitable development). The remainder of this section will examine events in Hong Kong.

Hong Kong: coping with rapid urban growth

Effectively the whole of the Crown Colony of Hong Kong constitutes a city-region. The main metropolitan area covers only about 50 square kilometres of the total of around 1,000 square kilometres, and until recently the remaining parts of the colony comprised sparsely used upland and/or island areas, together with some intensive agriculture.

The movement of population and industry out of the old metropolitan area around Victoria Harbour did not begin until the decision was taken in the mid-1950s to invest in a large-scale, low-cost housing programme. For the next 20 years there was intensive housing development, not in the existing urban area where sites were too small and land acquisition was complex and costly, but in the area known as New Kowloon between the old tenements and the rugged Kowloon Hills (Figure 8.2).

As Figure 8.3 clearly indicates, this had a substantial impact on population redistribution within the colony, with 2.2 million people being accommodated in public housing by 1973 and most of these living in New Kowloon. This was not a smooth process, however, as many families objected to being moved to what was then the periphery of the city without adequate and cheap transport facilities. Such attitudes changed as more jobs were provided in the peripheral areas, particularly in the 'new towns' of Tsuen Wan and Kwum Tong. These two areas were never really new towns in the British sense. However, Tsuen Wan has retained the nomenclature and has expanded considerably over the years, now encompassing nearby Tsing Yi island.

Until the mid-1970s, Hong Kong's expansion had been contained within the harbour area to the south of the Kowloon Hills. The principal impact on the New Territories, which comprised the bulk of the outer urban-region, was to change and intensify agricultural land use. In 1973, however, a new ten-year housing programme was unveiled in which homes for 1.8 million people were to be constructed by the newly created Housing Department. It was estimated that only 15 per cent of this total could be accommodated within the existing built-up areas and the New Territories were to take the remainder.

The housing target for the New Territories was carefully planned to occur within an expanded programme of new towns. The three largest of these new towns are Tsuen Wan, now virtually contiguous to the metropolitan area, Sha Tin and Tsuen Mun (Figure 8.2). They will eventually contain about 70 per cent of the total planned new town population of 3 million.

The new town programme was organized around the concepts of 'balanced development' and 'self-containedness', both of which were borrowed from the British planning system and which have a tendency to mean all things to all people. Certainly one of the objectives of 'balance' was to decentralize people from the heavily congested central areas and this has undoubtedly been achieved, with the population housed in the old metropolitan areas of Kowloon and Hong Kong Island falling rapidly between 1971 and 1986.

In terms of internal self-containedness and balance, the new towns have proved rather less satisfactory. Few people and fewer firms actually want to move to the New Territories. The big incentive for individual families has therefore been the low rents (and high standard) of the government accommodation on offer. Moreover, the waiting time for a flat is less than half of the nine years necessary for a unit in the main metropolitan area.

As a result, the new towns tend to be dominated by public housing (70 per cent) and the less skilled workers. Only with the advent of middle-income public housing has the class bias begun to 'balance' out. Even so, employment opportunities in the new towns are more limited than was originally hoped and most new town residents still commute to their work in the metropolitan areas. This has necessitated huge expenditure on transportational improvements including road tunnels through the Kowloon Hills, new high-speed rail track and a light railway system. As the shift of population to the new towns continues, this commuting problem will become even more acute, particularly if Hong Kong continues to grow as a financial and service sector since most of these developments will be in the central city.

Transportational links are, indeed, one of the few ways in which planned integration between the inner and outer regions of Hong Kong have occurred. The development of the city-region seems to have taken place primarily to decant population and the concept of balance is something which has been (somewhat unsuccessfully) related to the internal composition of the new towns themselves. Little consideration appears to have been given to the notion of balance within the city-region as a whole, particularly to the impact of new towns and new industries on the environment of the New Territories. Case study E (Chapter 3) has already highlighted the problem of water pollution, and other equally serious environmental issues remain unresolved. In particular, small industrial establishments, many of them relating to refugee inflows, have sprung up all over the New Territories, as has illegal residential accommodation (Figure 8.5). The result is visual and physical pollution of an extensive nature.

Admittedly, the Hong Kong government has created a system of country parks in the New Territories, as much to protect water catchment areas as to provide recreational space. But the overall impression in the outer parts of the city-region is one of islands of planning control (the new towns, country parks and reservoirs) in a sea of developmental turmoil.

Furthermore, the prolonged emphasis on peripheral urban development, which began in the 1960s, has led to a persistent neglect of the inner city. In these densely populated tenement areas, redevelopment has been piecemeal and has largely been the result of private sector attempts to intensify residential densities (and profits) or to extend the central business district (and profits). Public sector initiatives have, for the most part, sought to facilitate such developments by creating more building space through land reclamation or by clearance and demolition of the worst tenements, or else by improving access. Needless to say, all of these changes are to the detriment of the quality of life of the low-income residents.

Somewhat belatedly, the mid-1980s' review of planning directions in Hong Kong has sought to redirect attention towards the inner city. However, little in the way of firm government commitment has yet been forthcoming and it is difficult to dispel the impression that this concern conceals a more fundamental desire to redevelop the existing airport and container port. Certainly many planners have criticized the new

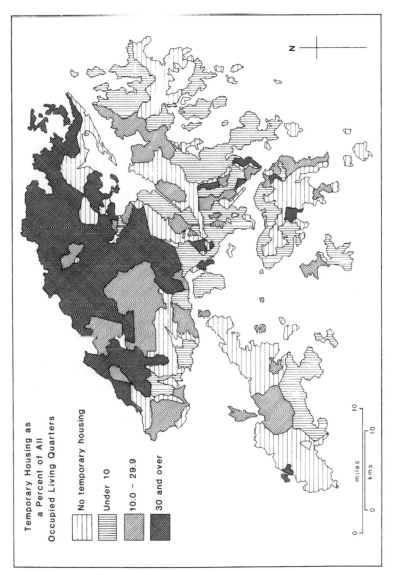

Temporary Housing as a Percent of All Occupied Living Quarters

No temporary housing

Under 10

10.0 – 29.9

30 and over

Figure 8.5 Hong Kong: extent of temporary accommodation

strategic plan for the colony for its perpetuation of the top-down approach, although the notion of strategic planning for the colony (or city-region) as a whole is to be welcomed.

It is very tempting for visitors to sophisticated Hong Kong, on seeing the spectacular city centre and the well-planned estates of the most recent new towns, to assume that they have seen the future for Pacific Asia. However, a closer examination of some of the social, environmental and spatial disparities which still exist might convince them otherwise. This may seem overly critical, since the majority of Hong Kong's residents seem content with their immediate life-style (if somewhat concerned for the future) and, moreover, the place makes money. But planning and structuring a city-region with profits and efficiency in mind must result in some people and places being bypassed, and to some these socio-spatial costs in Hong Kong are becoming increasingly unacceptable.

Key ideas

1 Although not the most urbanized region of the Third World, Pacific Asia has intensive urban development.
2 The levels and rates of urban growth in the region vary widely.
3 The urbanization process has been closely tied to economic growth. Some would argue that economic efficiency has outweighed social equality in this process.
4 Within the various elements of urban development, such as transport or housing, there is often conflict between the formal and informal sectors which the state tries to resolve in favour of the former.
5 This conflict is most acute in the urban periphery.

9
Gender and development: migration, urbanization and industrialization in Taiwan

The growth of interest in gender and development

Over the last two decades there has been a proliferation of interest in, and information on, gender and development. This was both a cause and a consequence of the establishment of the United Nations Decade for Women between 1975 and 1985. This decade saw little direct change in the subordination of women but it did help to promote greater awareness of the situation and the need for more information.

Even today, information on gender differences in the development process is patchy. Women are overlooked because they are 'economically inactive' in a formal sense. Their work is confined to the home, or to subsistence food production, or else they are involved in informal sector work, part-time work (often at home) or comprise unpaid family workers. None of these activities is exclusively female, so that under-recording affects the appreciation of male as well as female roles in the economy, but women tend in many developing countries (not all) to be more represented than men in those sectors of the economy about which we lack information.

The upsurge in interest in gender roles within capitalist development is not simply a consequence of the growth in feminism in developed economies. Women engaged in research within their own developing countries tend to display varied views on this question. Some see feminism as irrelevant and do not wish to separate gender issues from the broader aspects of underdevelopment with which they are concerned, such as poverty or multinational exploitation. Others, however, see feminism much more positively as a 'decolonization process' whereby the retrogressive images of women have been gradually stripped away. In addition to the influence of feminism, recent changes in the gender dimensions of development have also created more interest. In particular, the massive growth in the female labour force in certain economic activities and particular parts of the world, mostly notably the factories of Pacific Asia.

As a result of the past paucity of information on gender, the principal theories on development, whether from the political left or right, have been largely devoid of relevance in this context. Indeed, for most capitalist development strategists, the so-called trickle-down effect was supposed to be as appropriate for women as it was for the poor. In reality neither have benefited, despite the fact that both have been instrumental in accelerating the development process.

Table 9.1. Third World regional variations in gender status

| | Female activity rate | | Equality | |
	Rural	*Urban*	*Social*	*Economic*
Pacific Asia	high	high	low	high
South Asia	high	low	low	high
Africa	high	low	low	high
Latin America	very low	high	high	low
Middle East and North Africa	very low	low	low	low

Source: Adapted from Momsen and Townsend (1987)

Yet identifying the gender dimensions of underdevelopment is not easy. There are not only substantial differences between the major areas of the Third World (Table 9.1) but also between nations within those regions, as well as within individual countries themselves (between urban and rural areas, for example). So it is not surprising that Janet Momsen and Janet Townsend have observed 'the interrelationships of gender with race and class, core and periphery, rural and urban make for a very complex picture' but 'it must be accepted that the implications of gender in the study of geography are at least as important as the implications of any other social or economic factor which transforms society and space.' Our objective must be, therefore, not to study women as a separate issue in development but as part of other broader developmental processes. In this instance our framework will be migration, urbanization and the growth of multinational industry, primarily in the context of Taiwan.

The evolution of gender roles in development

It is still widely assumed that female status in traditional pre-capitalist societies was much more subordinate than it is today. However, this does not always seem to have been the case, with two principal factors being important in affecting the situation, viz. the predominant religion and the prevailing agricultural system.

Several major religious systems in Pacific Asia, particularly Buddhism and Confucianism, do allocate an inferior and subordinate role to women. In contrast, Islam gives equal rights to women, particularly over the inheritance of property, but cultural practices in Muslim societies have also confined women largely to the home. The major agricultural divisions within pre-colonial society also played an important role. In general the simpler, less technical systems, such as shifting cultivation or hunter–gatherer communities, tended to be those with a female contribution that was equal to, if not more important than, that of men. For example, in hunter–gatherer communities the basic subsistence food was gathered by the women, whilst men hunted for the more irregular luxury of meat.

In more agrarian, plough based societies the direct labour input of women was usually much reduced since technology (the plough) was monopolized by men. However, the intensive labour demands of paddy rice cultivation meant that the whole family was mobilized as a labour force. In general, however, the more complex the political economy, the more it was likely to be dominated by, and run for the benefit of, men. Unfortunately, one cannot generalize in the same way about systems in which women's labour was important, since cultural factors often restricted their social, economic and political power at any level within the community (Table 9.1).

Whatever the local situation was with regard to gender roles in pre-colonial Pacific Asia, colonialism wrought enormous changes, largely as a result of new labour demands and the introduction of new technologies. Overwhelmingly, these changes were directed at men. Men dominated the political economy of all the colonial powers so that the new patterns of cash cropping, wage labour and so on were imposed on or offered to men. New legal codes vested ownership or tenure of land and property with men; education was for men; and it was men who migrated to the new jobs in the colonial city (most urban domestic servants during colonial times were men).

Colonial capitalism was, therefore, highly patriarchal in nature. Men controlled resources and defined the productive and reproductive roles for women. In effect, the women of Pacific Asia were left in the traditional feudal world, it was the men of the region who were drawn into capitalism; women became subordinate to an already subordinate subsystem.

The irony of this situation is, of course, that women became even more essential to the survival of the household unit as men moved out of subsistence activities and into the colonial capitalist economy, either through rural wage-labour or by migrating to the cities. Women thus assumed an even greater role in domestic agriculture, growing the crops, selling any surplus, as well as undertaking traditional craft work (again for household use or commercial exchange). Many women even took on part-time employment too. All of which was in addition to normal domestic work of household reproduction.

The colonial legacy as far as women were concerned, therefore, was a heavily patriarchal system which has continued to affect the gender dimensions of the development process. Indeed, although it is difficult to generalize, the situation has worsened for many women. This is particularly true of the rural areas where the advance of technology, in all its forms, continues to endorse the assertion that technological change has usually been to the overall benefit of men.

However, this situation must not be overstressed. Many of the rural changes that have occurred since independence have disadvantaged all rural residents, not just women. Moreover, women themselves do not necessarily interpret events as a patriarchal conspiracy. Their primary concern is frequently with the poverty of the family and not with their status as women. What many want, in most instances, is simply access to better health, nutrition or educational opportunities, i.e. change from a class rather than a gender perspective.

The continued absence of change for the better in the rural areas, coupled with the growth of manufacturing employment in the cities, much of it open to women, has in recent years induced change in migrational patterns in Pacific Asia. Until relatively recently it was assumed that most female migrants were passive; dependently linked to the migration of their families, fathers or husbands. This has now changed and the study of the gender dimensions of migration has been identified as important for several reasons.

First, women represent a different human resource than men. Their rural economic roles are different so their migration will have repercussions previously unexperienced. Second, women are subject to different influences on their decision to migrate. Third, female migrants pose different problems in the city. For example, housing for single women is often in very short supply; married and single women tend to drift into different sectors of the urban labour market; more women in the labour market means pressure on child-care facilities, whether this is provided by the state or the family. Finally, female migration has important effects on household structure (the rise of the female-headed household) and on fertility. Both of these may have a considerable long-term impact.

Unfortunately, apart from emphasizing the fact that young, single women are now an equally important, and in some countries a more important, part of the migration stream, the limited aggregate data statistics available offer little in the way of explanation or discussion of impact. The opportunity to migrate and take on a major financial role must change female relationships within their families. This will be further affected by the improvement in the cost and speed of transportation which has permitted much greater contact between rural and urban areas than has been the case for migrants in the past. We cannot analyse such changes through census data; information must come from individual research studies. It is this approach that we will adopt later in this chapter.

However, before we move into a more specific type of study, it is necessary to consider some of the general themes related to gender changes in the contemporary urban economic structure. The important point to remember here is that the new characteristics of female migration to cities have, for the most part, been caused by radical changes in the nature of the world economy. In short, we cannot separate an examination of gender issues from the broader societal processes of which they are part. All of the major aspects of the changing world economy that we have already examined – the new international division of labour, growing multinational investment in the Third World, accelerated urbanization, increasing commercialization of agriculture and the like – all have gender implications.

There have been attempts to theorize how these broader changes articulate with gender and to explain what amounts to the sexual division of labour in the city – in particular the roles of women in the so-called formal and informal sectors. We will examine some of these theories and their principal arguments below, but it is worth noting at the outset that many women undertake work which falls into neither the formal nor the informal sector.

Women's work in the city: some theoretical ideas

Formal sector

Although the sexual division of labour is a common feature of most contemporary societies, it need not be accompanied by the subordination of either sex. However, the reality is that women have a subordinate role in which their work is either considered to be 'inferior' or 'secondary' (in the sense that they are not the main breadwinner) and is rewarded less; or else, as in the case of household or home reproduction, it is not rewarded at all.

Several models which have been put forward in the past to explain the subordinate and exploited role of women in the formal urban labour market have been linked to their alleged biological weaknesses as compared to men. The 'human capital' model, for example, alleged that women are less valuable to the capitalist system because they are less productive owing to the fact that they are neither as strong nor as skilled and are subject to a greater degree of absenteeism and labour turnover, i.e. they stay in jobs for shorter periods. Of course, many of these alleged shortcomings are themselves functions of broader societal attitudes towards women. Women are sometimes prone to short- or long-term absences from work because of the demands of their home and family, whilst the fact that some may offer fewer skills is a result of both inferior education and job discrimination.

Similar logic occurs within the other models of explanation. The 'institutional model', for example, suggests that there are two types of employment: the first is 'static', placing no great demand on skills and, therefore, permitting a greater degree of turnover and absenteeism. These are the jobs alleged suitable for women. The second category is 'dynamic' or progressive employment in which continuity is important to acquire skills and advancement. As you might suspect, these are supposedly career jobs for men, who are the principal long-term breadwinners for the family. A related theory, the 'overcrowding model', builds on the alleged employment limitations of women, in terms of their biology, aptitude and reliability, to suggest that women will tend to be pushed towards or crowded in those occupations that are more 'appropriate'. This causes frantic competition for relatively few jobs and holds down wages.

Most of the theories outlined above imply that as women are peripheral to the demands of urban employers, they will not be well represented in the formal urban labour market. However, as we have seen in previous chapters, this is not the case. Both female migration to the cities and increased absorption into wage labour has occurred on an increasing, if patchy, scale. We, therefore, have to seek alternative ideas as to what has happened.

In this context we cannot ignore the new international division of labour which has led to the mushrooming growth of factories in certain parts of the Third World and a corresponding demand for cheap labour. Much of the work is poorly paid and monotonous assembly labour, in which a high proportion of females is employed. Explanations for this from employers usually emphasize the manual dexterity of women but in reality it is because they comprise a more docile labour force. It is thought that most female factory workers are young, single women who want to work for a relatively short period before they get married, but this view is being challenged by new research which suggests quite a different family relationship in which the older females make a commitment to urban wage labour in order that their younger brothers and sisters may extend their education (see Case study P).

Opinions are divided too on the benefits to women of this new incorporation into the formal sector of the economy. Some feel that it is a positive step, bringing women out of the shadows of traditionalism into the modern world and giving them new cash incentives and new responsibilities. Others are more sceptical that these 'achievements' will lead to any real change for the better, pointing out the degree of stress, both physical and mental, to which women in multinational factories are subjected. In short, much female participation in the formal factory sector of the capitalist economy tends to be short lived, poorly paid and in few ways can be considered as comprising an improvement in status.

Case study P

Women's work patterns in Filipino cities

Migration patterns to cities in the Philippines are quite different to those in Taiwan and the rest of Asia. Migration over longer

Case study P *(continued)*

distances from rural to urban areas has been characteristic of Filipino women since the 1950s. However, throughout this period, until very recently, married women accompanying their husbands have been as numerous as single women. The explanation for this lies in the rural land tenure patterns which are similar to those in all former Spanish colonies and give rise to numerous small tenant farmers within huge estates. In these circumstances, there is little incentive for women to stay behind because the land does not belong to the family.

Again until the 1980s, migration was not as concentrated on the capital city as elsewhere in Pacific Asia, the reason being that, by and large, the prospects of employment and general quality of life were better in the smaller and intermediate cities. Overall, how-ever, those in remunerative employment worked in a variety of occupations but most were concentrated in the service sector and domestic service in particular.

In the last decade or so, multinational manufacturing has come to the Philippines, largely attracted by government incentive packages located in export processing zones. The country has few natural resources and poor energy reserves, so that almost all the components have to be imported. Clearly, cheap labour for assembly is the big attraction and former President Marcos stated his determination to retain this advantage:

Our country now has one of the lowest average wage levels in this part of the world ... [and we] intend to see to it that our export programme is not placed in jeopardy ... by a rapid rise in the general wage level.

As yet, however, the contribution of export industries to the economy and to the urban workforce is relatively limited. Many women are associated with urban industry only on a part-time, casual or outwork basis, and continue to rely on domestic service or petty commodity activities for their main sources of cash. Indeed, domestic service employment abroad has become almost as popular as factory work for young, single women. Almost

Case study P *(continued)*

30,000 work in Hong Kong alone and virtually take over its city centre on Sundays.

What follows is a composite profile of women and the urban economy in Manila derived from unpublished reports by Maita Gomez and Rosalinda Pineda-Ofrenea to the Irene Conference on Women and the Informal Sector held in Holland in 1985.

Maria is a widow aged 37. She has four children; two daughters aged 20 and 15, and two sons aged 16 and 12. She came to Manila when she was 6 years old with her parents who were both fish-paste vendors. Maria attended school until she was 11 but then had to leave because her parents could not afford to keep all their children at school and decided that the boys made the best investment.

For a while Maria looked after her younger brothers and sisters in their squatter house in the huge Tondo settlement; gradually she began to help her parents and it was whilst selling the fish-paste that she met her husband Eduardo who was already a vendor of sweets, newspapers and candies from a more or less permanent pitch. In the years that followed their common-law marriage, she not only helped her husband in his business, but also had nine pregnancies, five of which ended in miscarriages. Her husband died two years ago and she took over the business herself. As a result, her working day is now extremely demanding.

It begins at 4 am when she wakes and cooks breakfast for her family, usually any leftovers from the night before, and then goes to the public market where she buys the variety of goods that she sells. Her stall is near the entrance to the main Port Authority harbour area in Tondo, where her patrons are the dockworkers themselves, many of whom also live in the Tondo and know Maria and her family.

Maria sells from 7 am to 11 am, from 2 pm to 5 pm and most evenings to 11 pm In the intervening periods she prepares her family's meals and whilst she is at home her younger daughter looks after the stall. Maria's net earnings, despite all these long hours, amount only to £3 or £4 per day. She cannot afford to pay the full licence fee for vending which costs £60, so she pays the police about 80 pence per month instead.

Case study P *(continued)*

To ensure that her sons' education continued, Maria took her younger daughter out of school soon after her husband died. Both boys are doing well and Maria is confident that her eldest son, Ramon, will stand a good chance of obtaining a job at a foreign factory in the export processing zone when he leaves school later this year. If he does, the family's financial situation should improve dramatically but until then it is sustained by the work of the two girls.

Rosa, the elder daughter, has been working in domestic service now for six or seven years. She works for a middle-class family, where the head of the household is a bank clerk. She began as a laundress and general cleaner but is now the family cook and gets on very well with her employers. Her monthly salary is £70 per month, near the top end of the scale, and although she lives at home, she eats at her employer's and sometimes is able to bring home some food for her own family. Rosa knows that she is good at her job, that is why her employers pay her reasonably well, and she has ambitions to capitalize on this by going abroad to work. She knows that government regulations will make sure that 75 per cent of her salary is remitted back home but her real goal is to marry abroad. To this end she has advertised in a newspaper which circulates amongst potential employers and husbands in Australia. Maria knows of her plans and fully supports them, for it will mean further financial security for the family if her daughter is successful. However, Rosa is competing against many professionally qualified women in the overseas job market and her chances of success depend on her looks. She has invested a lot of money in a good photograph for her advertisement and is hoping for the best.

Maria, the younger daughter, works mainly at home. She does much of the housework, whilst her mother is at her stall, and in addition she works from time to time for a nearby government factory. Work is only available at certain times of the year and Maria Jr. is dependent on the agent to put the sewing her way. However, a few of her mother's sweets for the agent's children has proved to be a good investment in this respect. The factory is locally owned and the pay is poor, but Maria Jr. hopes that the

Case study P *(continued)*

experience she gains will help her land a full-time job with a bigger company at some point in the future.

In short, times are hard at present but hopes are high that in the future the investment in the boys' education will provide financial security. All members of the family are struggling against a system that exploits poor people but the women are doubly disadvantaged since they have all made substantial sacrifices in order to fulfil the expectations and demands of a strongly patriarchal society.

The petty commodity or informal sector

Marginalization theory suggests that women are squeezed out of formal sector employment and into the petty commodity or informal sector where they work long hours for little remuneration. Whilst there is some truth in this assertion, we have already noted that women are increasingly gaining access to formal sector work. Does this mean that petty commodity employment is no longer the prerogative of women, if it ever was?

The main types of employment within this range of activities are outlined below.

Casual wage labour which is taken on for short- or long-term periods with little job security, lower rates of pay and no side-benefits. The gender of the labour in this category tends to follow that of the permanent labour force but casual work or illegal employment does tend to characterize industries in which women predominate, such as textiles. One report from the Philippines by Maita Gomez, a political activist, cited a company with four garment factories, of which 95 per cent of the 3,200 workforce was female. Of this total 1,700 were casuals, 600 of whom worked in one factory where they comprised the entire labour force. If there was labour trouble at any of the other factories, work was simply switched to the casual location where more workers were recruited.

Subcontracted work or outwork is undertaken outside the factory, usually in the home. Again this is used when demand is high and the factory owner avoids the expense of excessive capital investment; benefiting also by paying lower wages. Indeed, the part-time worker is often totally dependent on the capitalist enterprise for supply of materials, rent of machines (if necessary) and purchase of the finished product. Women predominate in this kind of work, even in developed countries.

Self-employed workers are those who are most numerous in the petty commodity sector. They cover an enormous range of activities from cooked-food hawking to scavenging recyclable materials from rubbish dumps. Although these activities often encompass a large number of women, particularly single parents, not all need be thought of as poor. As with the informal sector as a whole, a range of incomes are earned but usually after long hours of work, coupled with domestic chores too.

Finally there is *unpaid family labour* which is perhaps the most insidious form of female exploitation since no direct remuneration is received for work undertaken on an enterprise that is usually run by a male relative.

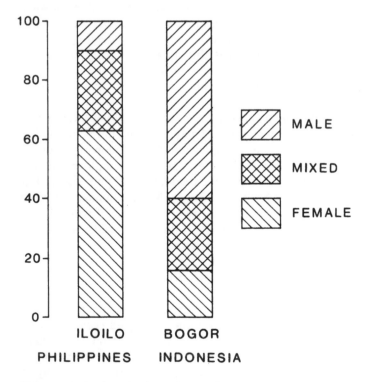

Figure 9.1 Cooked-food vendor gender

Even though we can generalize to a certain extent about the above categories, there is little gender consistency about petty commodity production, even within Pacific Asia. Figure 9.1 for example, reveals that one of the better-paid activities, cooked-food vending, has quite different gender profiles in the Philippines and Indonesian cities studied. Part of the explanation for the difference may lie in the fact that Bogor is situated in one of the stronger Islamic areas of Java, but the difference is still very marked, more so when it is realized that many men only entered into cooked-food hawking in the Philippines when their wives had established the viability of the enterprise. It must be noted also that these data only cover the vending of cooked foods; the purchase of the food and its preparation (which may or may not be separately remunerated) is almost entirely undertaken by women.

However, women's petty commodity activities ought not to be considered as secondary work. Not only are these activities vital to the profitability of formal capitalism, they are usually crucial to the household unit. For many families, the money earned through petty commodity activities is the basis of its budget; casual male employment has a more occasional value. This is not too dissimilar to gender roles in hunter–gatherer communities.

Domestic activities

Women who work in and around the house are often officially categorized as 'economically inactive'. Nothing could be further from the truth. Although this is a contentious area of debate amongst radical and feminist theoreticians, most agree that domestic work is part of the broader role of the reproduction of labour that capitalism allocates to women. In order for capitalism to continue, labour must be produced, reared and socialized into its role within the capitalist system. This task is largely undertaken in the home, exclusively so prior to schooling.

It has been asked, in this context, that if women's role in the home is so essential to capitalism, why are so many being drawn into the wage economy? The answer is, of course, that capitalism makes its most direct demands on young, single women. At other times, married women are expected to retain their domestic position, hence the variety of part-time or petty commodity activities which enable them to combine both roles.

The situation is more complex than this, however, since women also often undertake a variety of tasks which are not directly remunerative but are of indirect economic benefit to the family. These activities are

Table 9.2 Pacific Asia: changing economic participation by gender

	Urban % of total pop.	FLFP rate		Female % of total urban employment		Relative growth of male–female employment in the 1970s					
						Manufacturing		Trade		Services	
		1970	1980	1970	1980	% female employment in sector	Av. ann. growth	% female employment in sector	An. ann. growth	% female employment in sector	Av. ann. growth
Hong Kong	92	42.8	49.5	33.7	35.5	54.1	4.4	22.4	13.7	18.1	2.6
Indonesia	22	25.5	27.7	26.4	28.8	24.1	5.3	41.1	5.8	32.2	8.7
South Korea	65	26.3	20.4	35.1	31.1	40.8	6.9	34.4	5.7	16.6	2.6
Malaysia	32	28.2	45.0	26.2	–	35.4	16.6	20.6	14.5	39.1	9.6
Philippines	40	33.7	–	37.4	–	23.6	-7.1	22.3	1.4	50.2	1.0
Singapore	100	29.5	44.2	23.6	34.4	40.8	11.4	21.4	10.0	30.9	5.7
Taiwan	67	26.7	34.2	17.1	32.0	48.1	16.2	19.9	13.6	27.3	8.2
Thailand	17	45.1	54.4	39.2	43.4	23.9	7.2	37.7	7.1	34.0	6.6
Japan	76	47.7	44.0								
USA	74	41.4	44.3								

Source: G. Jones (1983) 'Economic growth and changing female employment structure in the cities of Southeast and East Asia', Department of Demography, Australian National University, Canberra

many and varied, ranging from the collection of fuelwood to subsistence gardening (legal or illegal), maintaining the family habitat (and therefore property values) or making/repairing clothes. Obviously, the gradation of activities between the petty commodity and domestic sectors is as difficult to conceptualize as those which span the formal–informal categories. Clearly generalized theories can take our discussion only so far, we now need to examine in detail the realities within the Pacific Asia region, and within Taiwan in particular.

Women in the workforce in Pacific Asia

By Third-World standards, female labour force participation rates (LFPR) in Pacific Asia are relatively high (Table 9.2), being compatible with those of many western countries. This is even more impressive given the differences in the extent of urbanization, which tends to be lower in Pacific Asia than in the Middle East or Latin America; although within the region the level of urbanization and LFPR do not correlate particularly well, with the two least-urbanized countries, Indonesia and Thailand, exhibiting the greatest contrast in female LFPRs.

Clearly, other factors are at work in influencing national female LFPRs, in the case of Indonesia it may well be Islamic conservatism, although this is not borne out by Malaysia. On the other hand, it was considered that for many years female LFPRs in Pacific Asia were positively correlated with the predominance of ethnic Chinese in the urban population. There is still some truth in this explanation but other ethnic groups are beginning to catch up rapidly.

The fact that cultural differences are important in affecting regional characteristics can be seen in Figure 9.2 which indicates female LFPR by age. These graphs clearly reveal the importance of young, presumably single women in the workforce of rapidly industrializing countries. The sharp peaks of Singapore, Hong Kong and Seoul and, to a lesser extent, Taipei and urban Malaysia contrast markedly with the smoother age profiles of the female workforce in Manila, Bangkok and Jakarta. In all cases, this tendency towards a younger female workforce has sharpened through time.

In general, married women are less involved in the labour force than single women, but Figure 9.2 indicates a return to the formal workforce by older women, presumably following marriage and motherhood, in those cities where economic growth is most rapid. What the graphs do

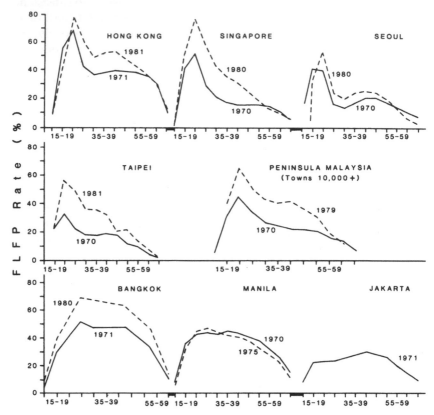

Figure 9.2 Pacific Asia: urban female labour force participation rates

not reveal, however, is the large number of older women in outwork or petty commodity activities which tend not to be recorded.

Female LFPRs are also much higher for recent migrants into the city, particularly for single, young women. However, explanation of the rise in migration and labour force participation is more difficult. It is not enough to point to the greater availability of factory jobs (some traditional female urban occupations such as domestic service are declining); the erosion of parental control and traditional attitudes about the family (such as an idle wife being indicative of a husband's ability to provide for her); and the improving level of women's education, which are only a few of the many other factors involved. As a result, it is often difficult to distinguish between women who move into urban employment because their family needs the money, and those who, on the other hand, are seeking to assert their independence.

Since the 1960s, women's share of urban employment has been increasing steadily, spectacularly in some instances (Table 9.2), although there are clearly differences among individual countries. Manufacturing has risen considerably in importance, particularly in the Pacific Asian NICs where it is now the leading employer of women. It is worth noting in this context, that manufacturing work is not necessarily dominated by recent migrants. Presumably, residence in the city builds up the necessary contacts through which this work is obtained. It must also be remembered that not all manufacturing comprises multinational factory work; also included in this category are more traditional occupations of handicrafts, rattan weaving and the like (see Case study Q).

Within all of the Pacific Asian states, female employment in manufacturing does tend to be more concentrated than that of men into a narrow range of occupations. For example, one Malaysian study revealed that whilst men were involved in 74 occupations, women were in only 14. In Taiwan, one-third of the female urban workforce are in textiles and clothing, one-fifth in electrical or electronics industries. By far the great majority of these women are young and relatively uneducated, indeed many are illiterate. Perhaps not surprisingly, women tend to 'sediment' to the bottom of the wage scales; even within similar occupations to men, they receive lower wages.

Although manufacturing employment is undoubtedly becoming increasingly significant, in many cities the more 'traditional' female areas of occupation in retailing, office work and domestic service remain very important. The last of these was usually the most common first job in the city for many female migrants but this has declined in importance during the last two decades because pay was poor and working conditions unregulated. Even factory work is now more popular. As with manufacturing, these data have to be treated with some caution as they mask a shift from the more traditional trade and service occupations in small stores, market places, etc. to jobs in the new offices, department stores and fast-food restaurants, which reflect the growing western character of the central business districts of most Pacific Asian cities. Much of this has been intermingled with the growth of the tourist industry and its associated activities. Many young women are thus involved in a range of hotels, restaurants and other 'hospitality' services, such as bars, massage parlours and the like (for further discussion see John Lea's book on *Tourism and Development in the Third World* in this series).

Bearing in mind that this brief review of employment data is confined to the formal or capitalist sector, several important trends emerge from the analysis. First, that women have played a substantial role in the urban economic growth of Pacific Asia over the last 20 years. Second, that their rate of incorporation into the economy, particularly in the relevant export industries, continues to be faster than that of men. Third, that although the range of female occupations is generally widening, women are still heavily concentrated in certain areas of employment. Finally, that women continue to be underpaid overall and in relation to men. We will now examine the extent to which this is borne out in our case study of Taipei. We will begin with a brief background note on Taiwan and its capital city.

Case study Q

From traditionalism to socialism: women and rapid change in Laos

Laos is a small, land-locked country of 3½ million people. It is probably the least developed and least known country in Pacific Asia. It was colonized by France in the late nineteenth century but constituted a neglected backwater of the French Indo-Chinese Union. The departure of the French in the early 1950s saw Laos plunged into the same revolutionary conflict that engulfed Vietnam and Cambodia. For 20 years the communist Pathet Lao fought the royalists backed by the United States. The northern strongholds of the Pathet Lao were bombed incessantly whilst US aid propped up the royalists. Per head of population, Laos received the heaviest bombing and most aid of any country – two extremes which virtually destroyed the traditional economy. A self-sufficient economy became a food importer. Moreover, more than a quarter of the population became refugees, with the principal shift being from the communist-controlled northern provinces to the outskirts of Vientiane, the capital. Nor did peace in 1975 bring total stability: many Lao resented the heavy-handed socialist restructuring and by the mid-1980s, 10 per cent of the population had fled to Thailand as refugees.

Case study Q *(continued)*

Phonetong village was established in 1963 on the periphery of Vientiane by refugees from the northern province of Sam Nuea. For the women of the village the move was economically and socially traumatic. Lacking any skills for urban employment, including literacy, the women remained at home undertaking domestic work and gardening. However, in Sam Nuea they had been renowned for their weaving skills in traditional materials and designs – skills which were essential to their marriage prospects.

In Vientiane this weaving drastically declined. The raw silk and natural dyes were difficult to obtain, there was competition from artificial materials, and most people preferred western clothes and styles anyway. The little traditional weaving that remained was for small-scale private use. Thus geographical relocation had drastically reduced the women's contribution to the household economy, making it dependent on the earnings of men in an erratic urban economy.

This situation changed when Laos fell to the communists in 1975. US aid and Thai consumer trade ceased and the urban economy was undermined. Male employment dwindled, wages were cut and many men were sent off for re-education. As a result, women emerged again as the leading producers, this time of domestic subsistence production, mostly vegetables and small livestock. With the shrinkage of Thai consumer goods, including synthetic materials, and the new government's disapproval of western dress, traditional weaving also began to re-emerge. One enterprising woman, Komaly, decided to establish a cooperative.

Initially this began as an informal work team with Komaly purchasing materials, acquiring small contracts and selling to local merchants. Soon, however, she sought to formalize this arrangement within the socialist development goals of the new government. However, a cooperative demands capital input on the part of its members and most of the village were very poor. Eventually eight women joined Komaly and her sister Somalay, who acted as administrator/accountant, to form the new Phonetong Cooperative. Significantly, almost all were heads of new households and this was the only way to expand their earning capacity.

In 1980, a year after the commencement, the cooperative

Case study Q *(continued)*

received official recognition and subsidies, and orders became easier to obtain. Komaly found it easy to subcontract out to part-time workers and by 1985 the cooperative had grown to 60 full-time workers and 600–700 part-time workers. Most of the full-time workers are now younger women, whereas the part-timers are usually restricted to work in their homes because of domestic commitments, particularly children. However, some have full-time jobs and work in the evenings and weekends to pick up extra money. Only about 10 per cent of the workforce is male, they normally undertake the heavier work such as carpentry, packing and loading.

Wages for full-time workers amount to US$10–15 per month (in real terms) plus a range of welfare and other benefits. Remuneration is not large but is comparable to that of civil servants or factory workers. Part-time workers receive appropriate piece-rates but lack the welfare benefits. The great majority of the women spend their earnings on basics for the family such as food or clothing, only a few of the single girls spend any money on themselves. However, the women in the cooperative do not only work to supplement the household budget, valuable as that is, they are also conscious of the fact that the government wants more women to enter into production.

Despite this increased role in the workforce, all of the women continue to undertake all domestic work too. The concept of men sharing household work seems to be quite alien to both men and women. Thus the expanded economic role for women has assisted the household without changing its fundamental social structure. One reason why the women can cope with these dual labour demands seems to be that the cooperative is organized by women and operates within a flexibility designed to recognize the alternative pressures that may be placed on its workers by children, family events and so on. It provides a clear illustration of women's adaptability to rapidly and drastically changing circumstances.

(This case study is based on a summary of a much larger report on Phonetong by Ng Shui Meng in Noeleen Heyzer's edited book *Daughters in Industry*, published by the Asian and Pacific Development Centre, Kuala Lumpur, 1988.)

Economic development and urbanization in Taipei

Prior to 1895, Taiwan (or Formosa as it was then known) was a peripheral province of the Chinese empire, populated by fishermen and farmers, many of whom were of Malay origin. At the conclusion of the Sino-Japanese War in that year, the island was ceded to Japan, a status which it retained for the next 50 years. During this period the Japanese systematically developed its colony, reorganizing its agriculture along commercial lines and settling in large numbers into expanded cities, particularly the capital of Taipei and the new ports of Kaohsiung and Keelung (Figure 9.3). The economy was heavily structured around the production and processing of sugar cane and pineapples, and 90 per cent of all trade was with Japan.

Figure 9.3 Taiwan: principal urban centres

The great majority of the Japanese were city dwellers and comprised about one-quarter of the total urban population. However, although the towns grew rapidly under Japanese colonialism, so did the rural population, so that by 1940 the urban population was still only 15 per cent of the total. Most of the urbanization was initially centred on Taipei, which was substantially rebuilt and became something of a primate city. Later expansion elsewhere reduced this dominance so that between 1920 and 1940 Taipei's share of the total urban population halved from 40 per cent to 20 per cent.

In 1945 Chinee sovereignty was restored to Taiwan but China itself was riven by civil war between the nationalists, led by Chiang Kai-shek, and the communists, under Mao Tse-tung. The nationalists lost and in 1949 the remnants of the army and supporters fled to and occupied Taiwan, where they slowly asserted their control and independence with massive backing from the United States.

Gradually the economy took shape and expanded. Many of the nationalist refugees were former city dwellers and were skilled workers or entrepreneurs. All favoured a *laissez-faire* capitalism and an urban-based industrial economy began to emerge in the 1950s (see Chapter 7). Meanwhile, land reform increased small-scale ownership and helped retain rural population on the land but at the same time drove domestic investment capital into the cities, where it joined with state capital and foreign aid to provide the basis for economic growth.

By the 1960s a 'surplus' rural population had built up and the government decided to shift to export-oriented industrialization. Export processing zones were set up in several of the major cities, offering inducements to foreign firms. The result was an even more rapid rate of urban growth, much of which was of indigenous Taiwanese. Effective birth control programmes helped to cut the rate of increase in the 1970s, as did planning legislation, but a substantial amount of suburban expansion has occurred over the last 20 years.

Much of the migrant movement has been of women, and whilst there has not been the dominance of female migrants that we have noted elsewhere in Pacific Asia, there has been a massive growth in female representation in the manufacturing sector (Table 9.3). Free education (for nine years) has made the female migrant labour force one of the most educated in the world and foreign companies have been keen to invest and expand in Taiwan because of the availability of the labour, the range of government incentives and the substantial amount of local capital (multinational firms usually raise most of their investment capital in the country where they construct their factory).

Table 9.3 Taiwan: gender and employment (%)

	1970 M	1970 F	1975 M	1975 F	1980 M	1980 F
Largest cities*						
Agriculture	10.9	17.0	8.6	10.3	7.9	8.7
Manufacturing	25.3	23.3	29.6	31.7	31.8	33.8
Commerce	16.0	–	14.5	17.5	16.7	22.3
Transport	8.3	59.7	10.5	4.1	10.6	4.0
Services	39.2	–	36.7	36.2	32.8	30.5
Intermediate cities						
Agriculture	15.9	16.3	14.4	13.8	12.2	10.9
Manufacturing	22.8	29.2	30.5	38.9	35.2	43.7
Commerce	16.1	–	12.9	15.6	14.0	18.6
Transport	5.4	54.3	7.2	2.9	7.5	2.4
Services	39.8	–	34.9	28.7	31.1	24.4

Note: * Taipei, Kaohsiung, Taichung, Tainan, Keelung
Source: Liu (1980)

The decision to end martial law in 1987 proved a particular spur in this respect since it confirmed Taiwan's political stability. As a result, overseas investments doubled over the previous year, overwhelmingly in manufacturing enterprises (see Figure 9.4 and note the amount of Hong Kong capital searching for long-time investment security).

Taipei, in particular, has been a magnet for investment and for migrants. Almost half of those who left the countryside in the 1970s moved to Greater Taipei. Currently three-quarters of the population added to the capital each year are of rural origin. Not surprisingly, as in so many other Asian countries, a special capital city-region of 272 square kilometres was created and the population grew from 0.7 million in 1955 to 2.7 million by the mid-1980s.

Detailed information on development in Taiwan and Taipei is not easy to obtain, given the World Bank's refusal to acknowledge the country's existence. However, some individual studies have been published and the final section of this chapter, a review of female migration into Taipei, draws primarily on the work of Nora Huang (see Fawcett *et al.*, 1984).

Female migration to Taipei

In contrast to many other parts of Pacific Asia, there is not widespread poverty in the rural areas of Taiwan, largely as a result of land tenure reforms in the 1950s. However, farm life is usually hard and monotonous,

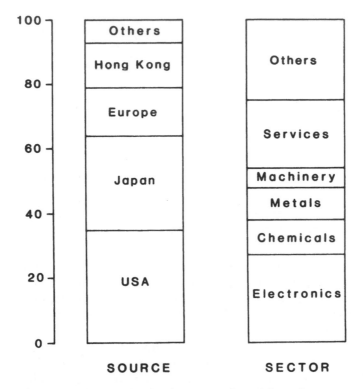

Figure 9.4 Taiwan: foreign investment by origin and sector

whilst the opportunities for alternative employment in small urban centres is quite limited. Not surprisingly, therefore, about half of the female migrants give economic reasons for moving from their village, although the desire to escape parental control and achieve both social and economic independence is also present. Another major reason for leaving the village is to gain further education beyond junior high school. After this level education is not free, so many migrants work their way through vocational schools, often taking low-pay or unskilled work as a means to the end of improved qualifications. Overall, very few female migrants (only 16 per cent) leave their village for passive or dependent reasons, simply moving with their family or husband. Even less (7 per cent) want to move because of a difficult or hard life in the village.

The great majority of the village migrants interviewed by Nora Huang

went to Taipei. Few thought of going to an alternative large city and 90
per cent of those who went to the capital moved directly there.
Clearly, improved communications and transport have made gradual
step-migration up the urban hierarchy somewhat obsolete. Few of the
women migrants had definite reasons for moving to the capital city, with
less than one-third being able to cite specific employment or study
objectives. Many simply had a rosy mental image of the capital
assembled from watching television soaps, reading magazines or visiting
and talking with friends who already lived in Taipei.

Certainly, the presence of relatives or friends in the place of destina-
tion is a very important factor. Although village girls may well be
asserting or seeking independence when they migrate, their families try
to ensure that the move itself takes places with certain prescribed
conditions, amongst which personal contacts in the city are paramount.

Indeed, these contacts prove very important for settling into Taipei,
with almost two-thirds having set up jobs before arriving in the city and
some 40 per cent staying with family or friends immediately after their
arrival. A similar proportion live in dormitory accommodation supplied
by their employers, usually fairly spartan and overcrowded.

Although the female LFPR in Taiwan is still not as high as in other
Pacific Asian NICs, there is much more of a concentration in the
manufacturing sector where women comprise almost half of the labour
force. Factory work is easy to obtain in Taipei and, indeed, suffers
shortages of skilled, trained labour. Yet working conditions for women
are often so unattractive that labour turnover is high and a factory job is
valued far less than white-collar work in an office or shop. However,
factory work is much preferred to domestic service. It might seem
sensible, however, in view of the labour turnover, for employers to
improve the longer-term prospects of their female workforce to induce
them to stay and build up their skills.

Unfortunately, such policies do not feature prominently in the
manufacturing sector and on average women earn only two-thirds the
wage of men. Much of the differential is due to occupational differences
but in some ways that is a consequence of anticipated employer
discrimination. However, even when women perform the same type of
work and have the same productivity, they still receive less. As a result,
many women earn what they regard as an acceptable wage only by
committing themselves to overtime – no doubt to the satisfaction and
profit of their employers. Little wonder then, that the money earned
from living in the city is not seen as the most important reward from

migration. Acquiring education, getting married and experiencing city life are all rated more highly.

For many female migrants rural ties remain strong, particularly whilst they remain single. In Nora Huang's surveys all of her respondents returned at some time during the year, with 45 per cent returning up to five times and 30 per cent six times or more. Most visits are linked to family events or holidays, especially at New Year when many women change jobs, but a sizeable proportion still return to help with rice harvests in those areas where migration has resulted in seasonal labour shortages.

Just under two-thirds of the female migrants are able to send money back to their village. Most send between US$30–80 which amounts to about one-fifth to one-quarter of their monthly incomes. This is a sizeable proportion of their wage given that they live in a high-cost city but, as Case study P indicates, many female migrants fund the education of their younger brothers and sisters.

In summary, although the role of women in economic development in Taiwan is broadly similar to that in other Pacific Asian countries, there are some specific local features. The push factors related to rural poverty and hardship are less marked and urban employment opportunities, at least in Taipei, appear to be relatively abundant. Despite the retention of rural ties, the move into the city, once made, tends to become permanent, particularly after marriage. Because of the availability of work in Taipei, more married women return to the workforce than in most other countries in the region.

All individual cities and states have their local characteristics, but fundamentally the role of the female migrant in Taipei is the same as it is in Singapore, Seoul or any other city in Pacific Asia – that of a cheap, docile pool of highly exploited labour. Given the greater educational opportunities that are opening up, particularly in Taipei, this situation may not persist. Nora Huang points out 'the perception of women workers as unskilled, patient and tolerant, and easily available, is now *passé* in Taiwan'. What is now needed is some form of labour union to translate wishes into reality.

Key ideas

1 The status of women was not always subordinate to men in pre-capitalist societies.
2 The advent of colonial capitalism favoured men.

3 In recent years more women have been incorporated to the work-force, primarily in factories.
4 Female labour force participation rates are currently high in Pacific Asia, but not solely in manufacturing.
5 Women are also important in the informal or petty commodity sector.
6 Despite these changes, women are still disadvantaged economically, socially and politically compared to men.

Further reading and review questions

Chapter 1

1 Why is the region under discussion now called Pacific Asia?
2 What is the nature of the trading links between Pacific Asia and the rest of the world?
3 Why are the human resources of the region so important?
4 Why are there variations in the standard of living in Pacific Asia?

Further reading

Far Eastern Economic Review (1990) *Yearbook 1990*, Hong Kong.
World Bank (1989) *World Development Report*, Washington DC.
World Resources Institute (1987) *World Resources 1987*, New York: Basic Books.

Chapter 2

1 What were the major phases of colonialism in Pacific Asia? Were they the same as in Latin America or Africa?
2 What were the reasons for the change in colonialism in the mid-nineteenth century?
3 Are there any films or television programmes you have seen, or novels you have read, that convey a strong colonial image? Which of the phases of colonialism do they represent?
4 Why has political independence not brought economic independence to many parts of Pacific Asia?

Further reading

Betts, R. (1985) *Uncertain Dimensions: Western Overseas Empires in the Twentieth Century*, Oxford: Oxford University Press.
Drakakis-Smith, D. (1987) *The Third World City*, London: Methuen.
Dixon, C. (1991) *Southeast Asia and the World Economy*, Oxford: Basil Blackwell.
Kirk, W. (1990) 'Southeast Asia in the colonial period' in D. Dwyer (ed.) *Southeast Asian Development: Geographical Perspectives*, London: Longman.
Osborne, M. (1985) *Southeast Asia: An Illustrated History*, Sydney: Allen & Unwin.

Chapter 3

1 Why are forests in Pacific Asia under severe threat at present (you may find Avijit Gupta's book useful in answering this question)?
2 Why are some mining workers so exploited and what can they do about it?
3 What is the difference between undernutrition and malnutrition? Has the nutritional problem in Pacific Asia improved in recent years and why (not)?
4 Why is education often regarded as the key to the improvement of human resource potential?
5 What is primary health care and why is it regarded as so important in current development programmes?

Further reading

Brown, M. (1972) 'Bougainville pays in copper', *Geographical Magazine* 44 (11): 734–6.
Forbes, D. (1982) 'Energy imperialism and the new international division of resources: the case of Indonesia', *Tijdschrift voor Economisch en Social Geografie* 73 (2).
Gupta, A. (1988) *Ecology and Development in the Third World*, London: Routledge.
Mountjoy, A. (1984) 'Core-periphery, government and multinationals: a Papua New Guinea example', *Geography* 69 (3): 234–43.
Soussan, J. (1988) *Primary Resources and Energy in the Third World*, London: Routledge.
World Bank (1981) *World Development Report*, Washington DC.

Chapter 4

1 Many observers have alleged that there is urban bias in Third World development. What evidence does this chapter reveal for and against this argument?
2 What were the reasons why the green revolution did not bring benefits to the poorer farmers? Why do governments still persist with this strategy?

3 Use the evidence presented in this chapter to evaluate the relative success of Thailand, South Korea and Vietnam in rural development.
4 Are rural and regional development the same?

Further reading

Dixon, C. (1990) *Rural Development in the Third World*, London: Routledge.
Grigg, D. (1986) 'World patterns of agricultural output', *Geography* 71 (3): 240–5.
Lea, D. and Chaudhri, D. P. (1983) *Rural Development and the State*, London: Methuen.
Thrift, N. (1987) 'Vietnam: geography of a socialist siege economy', *Geography* 72 (4): 340–4.
Wolf, E. (1986) 'Beyond the green revolution: new approaches for Third World agriculture', *Worldwatch Paper* 73.

Chapter 5

1 Why are population growth rates lower in Pacific Asia than in other parts of the Third World?
2 Why are rural population growth rates everywhere higher than in urban areas?
3 Have Malaysia and Singapore chosen to reverse their population control programmes for the same reasons?
4 What evidence is there for a 'brain drain' from Pacific Asia?
5 How has the Indonesian government sought to control urban population growth rates?

Further reading

Dwyer, D. (1987) 'New population policies in Malaysia and Singapore', *Geography* 72 (3): 248–50.
Forsyth, J. (1990) 'The Indonesian transmigration scheme', *Geography Review* 4 (2): 23–6.
Jellinek, L. (1988) 'The changing fortunes of a Jakarta street trader' in J. Gugler (ed.) *The Urbanization of the Third World*, Oxford: OUP.
Mansell Prothero, R. (1987) 'The people problem', *Geographical Magazine* 58 (2): 37–42.
Skeldon, R. (1986) 'Hong Kong and its hinterland', *Asian Geographer* 5 (1).
Ulack, R. and Leinbach, T. (1985) 'Migration and employment in urban Southeast Asia', *National Geographic Research* 1 (3): 310–37.

Chapter 6

1 How did ethnic mixing occur in the pre-colonial period?
2 How did colonial powers worsen and complicate the ethnic situation in Southeast Asia?

3 Has Malaysia or Burma been able to improve the ethnic situation since independence? If not, why not?

Further reading

Cho, G. (1990) *Malaysia*, London: Routledge.
Dixon, C. (1991) *Southeast Asia and the World Economy*, Oxford: Basil Blackwell.
Mehmet, O. (1986) *Development in Malaysia*, London: Croom Helm.
Steinberg, D. (1982) *Burma: A Socialist State of Southeast Asia*, Boulder, Colorado: Westview Press.

Chapter 7

1 Why have some countries industrialized more rapidly than others? Is it due to differences in resource endowment?
2 Has job creation in industry matched expectations? What alternative sources of employment are there in the cities?
3 Are the cultural characteristics of Pacific Asia really important in explaining industrial growth?
4 What is a *laissez-faire* economy? Are the four little tigers representative of this?
5 How did the four little tigers cope with the world recession of the mid-1980s?

Further reading

Berger, P. and Hsiao, H. H. (1988) *In Search of an East Asian Model of Development*, New Brunswick: Transaction.
Cline, W. (1982) 'Can the East Asian model of development be generalized?' *World Development* 10 (2): 81–90.
Dixon, C. (1984) 'The Far East after the boom years', *Geographical Magazine* 54 (2): 61–6.
Phillips, D. and Yeh, T. (1983) 'China experiments with modernization: the Shenzhen special economic zone', *Geography* 68 (4): 289–300.

Chapter 8

1 What are the reasons for the rapid rate of growth in Pacific Asia's cities?
2 Is this growth constant in all nations and for all types of cities?
3 Has urbanization followed economic growth or vice versa?
4 Has there been a social price to pay for urban economic success? Argue your case with reference to either housing or transport.
5 What is the role of the urban periphery in Pacific Asia?

Further reading

Dwyer, D. (1986) 'Land use and regional planning problems in the New Territories of Hong Kong', *Geographical Journal* 152 (6): 232–42.

Dwyer, D. and Sit, V. (1986) 'Small-scale industries and problems of urban and regional planning: a Hong Kong case study', *Third World Planning Review* 8 (2): 99–119.

Fuchs, R. J., Jones, G. W. and Pernia, E. M. (1987) *Urbanization and Urban Policies in Pacific Asia*, London: Westview Press. (Sixteen essays by various authors covering urban population growth, provision of services, transportation and decentralization.)

Jones, G. (1988) 'Urbanization trends in Southeast Asia', *Journal of Southeast Asian Studies* 19 (1): 137–54.

Chapter 9

1 What was the status of women in Pacific Asia in the pre-colonial period?

2 Why did this situation generally decline during the colonial period?

3 Has the shift in factory work improved the status of women to any extent?

4 Are there any differences between socialist and non-socialist states in terms of women's role in development?

5 In what ways have women been incorporated into urban growth other than through factory work?

Further reading

Fawcett, J., Khoo, S.-E. and Smith, P. (eds) (1984) *Women in the Cities of Asia*, Boulder, Colorado: Westview Press.

Hsu, A. and Pannell, C. (1982) 'Urbanization and residential structure in Taiwan', *Pacific Viewpoint* 23 (1): 22–52.

IBG Women and Geography Study Group (1984) *Geography and Gender*, London: Hutchinson.

Momsen, J. (1991) *Women and Development*, London: Routledge.

Momsen, J. and Townsend, J. (1987) *Geography and Gender in the Third World*, London: Hutchinson.

Townsend, J. and Townsend A. (1988) 'Teaching gender north–south', *Geography* 73 (3): 193–201.

Index